NUMBERS

AND THE MAKING OF US

Counting and the Course of Human Cultures

CALEB EVERETT

HARVARD UNIVERSITY PRESS

CAMBRIDGE, MASSACHUSETTS · LONDON, ENGLAND · 2017

Library of Congress Cataloging-in-Publication Data
Names: Everett, Caleb, author.
Title: Numbers and the making of us : counting and the course
of human cultures / Caleb Everett.
Description: Cambridge, Massachusetts : Harvard University Press,
2017. | Includes bibliographical references and index.
Identifiers: LCCN 2016035929 | ISBN 9780674504431 (alk. paper)
Subjects: LCSH: Numeration—Cross-cultural studies. |
Counting—Cross-cultural studies. | Number concept.
Classification: LCC QA141 .E94 2017 | DDC 513.2/11—dc23
LC record available at https://lccn.loc.gov/2016035929

For Jamie and Jude, who have enriched my life
in uncountable ways

CONTENTS

NUMBERS AND THE MAKING OF US

Prologue

ON THE SUCCESS OF OUR SPECIES

Survival is not easy. If you have ever ventured into an environment that has not been molded by contemporary society, you probably appreciated this fact quickly. Trekking through some tropical jungle on your own, for instance, this point is impressed on you with some severity. Aside from the discomfort of the sultry air and associated sweating (a poor adaptation in places with stifling humidity), not to mention such concerns as the various bacteria, viruses, insects, and more sizeable species potentially preying on you, you will find that the mere acquisition of water and food is onerous or downright impossible. If you ever have the chance to follow indigenous jungle-dwellers through the pliable Amazonian undergrowth, you will, if you are at all like me, become acutely aware of just how rapidly your surroundings could consume you were it not for the knowledge of those you are following. Juliane Koepcke, who famously survived the disintegration of her airliner thousands of meters above the Peruvian forest in 1971, startled the world when she survived the crash and lasted for more than 9 days alone in the jungle. As the teenaged child of biologists who worked in Amazonia, her knowledge of the surrounding ecology saved her life. Yet she was still unable to procure food during the whole of her ordeal and was eventually saved by members of a local riverine culture.

Most people in her position, isolated in the jungle, do not survive. The same is true of those lost in other unfamiliar pristine ecologies. The history of ocean navigation is littered with stories of explorers who were forced to rely on the local expertise of indigenous communities when stranded in new habitats. Televised "reality"-based depictions of individuals surviving in the wild without outside help are generally only made possible because the "solitary" survivor being filmed is provided with essential tools and supported by a team of producers who have prepared them in various ways for the environment in which they are "abandoned" along with their well-provisioned film crew. Humbling though it may be, you or I would likely die within a matter of days or, if luckier, weeks if left isolated in most of the world's ecosystems.[1]

More surprisingly, individual members of indigenous cultures often struggle in environments they know well, if they are accidentally isolated. Getting lost under the forest canopy may be comparatively less hazardous for those native to tropical jungles, for example, but it can still be a treacherous affair. I know of tribe members in Amazonia who have become perilously lost not far from their home village, only to barely survive or, in some unfortunate cases, to perish. Such cases drive home an important and often overlooked point: human survival is contingent on knowledge stored in the repository of culture, accessed through linguistic means. Daily, we rely on knowledge that is not really our own but can be easily extracted from the minds of others and has, in many instances, been brutally and often randomly acquired over the course of millennia. Consider some examples from your own culture: you did not need to invent the car, or indoor heating, or the most efficient method for filleting a chicken breast—you inherited such technologies and behaviors. You modeled your actions after others and were constantly taught your behaviors, either formally or informally, via language. The bulk of our daily activities, including those related to fundamental processes, such as eating and sleeping, are entirely dependent on ideas that we absorbed from those around us, who in turn absorbed them from others. While certain needs are biologically determined, our approach to handling those needs is constructed by our native culture. Nearly every material and behavioral invention that facilitates your life, from the toothbrush to shaking hands, was innovated by another human or set of humans. When it comes to ideas, we inherit much more than we innovate. And the same could be said of members of cultures radically different from our own. Hunters in New Guinea do not need to invent bows and arrows as the need arises—they inherit that technology through teaching and mimicry. Each generation of any culture builds on the knowledge of the previous ones, often acquired through accidental discoveries that may have followed painful or deadly events. For instance, bows and arrows and other basic hunting implements were not invented in one fell swoop. They evolved over the course of centuries as hunters gradually came to realize the life-saving advantages of some forms of bows and arrows over others, for particular purposes.[2]

Our increasingly refined means of survival and adaptation are the result of a *cultural ratchet*. This term, popularized by Duke University psychologist and primatologist Michael Tomasello, refers to the fact that humans cooperatively lock in knowledge from one generation to the next, like the clicking of a ratchet. In other words, our species' success is due in large measure to individual members' ability to learn from and emulate the advantageous behavior of their predecessors and contemporaries in their community. What makes humans special is not simply that we are so smart, it is that we do not have to continually come up with new solutions to the same old problems. We know what has worked in the past, though we do not necessarily know why it has worked in the past. Just because you can reheat a burrito does not mean you have the slightest notion of how to design a microwave or the electrical grid that enables its usage.[3]

The importance of gradually acquired knowledge stored in the community, culturally reified but not housed in the mind of any one individual, crystallizes when we consider cases in which entire cultures have nearly gone extinct because some of their stored knowledge dissipated due to the death of individuals who served as crucial nodes in their community's knowledge network. In the case of the Polar Inuit of Northwest Greenland, population declined in the mid-nineteenth century after an epidemic killed several elders of the community. These elders were buried along with their tools and weapons, in accordance with local tradition, and the Inuits' ability to manufacture the tools and weapons in question was severely compromised. This and other knowledge loss subsequently impaired their efforts to hunt caribou and seals and to harvest cold-water fish. As a result, their population did not recover until about 40 years later, when contact with another Inuit group allowed for the restoration of the communal knowledge base. In the course of human history, other cultures have died off completely due to analogous degradations of their survival-related know-how or because of the loss of basic material technologies that could not be easily replicated.[4]

Such cases directly contravene the popular, some would say mythologized, notion that humans excel simply because we are inherently smarter than other species. It turns out that this idea is poorly supported. While we are obviously smarter than other species and do have a high encephalization quotient (large brains for our body size), in some ways our innate cognition is not as advanced as we once assumed. Many of our distinguishing intellectual attributes are not genetically hardwired but learned in culturally dependent ways. While natural selection has undoubtedly yielded remarkable human brains, what is really most striking about our species is what we have managed to do with those brains since the advent of culture. In this book I join the crescendoing chorus of anthropologists, linguists, psychologists, and other *ists* who emphasize this point. These scholars stress that culturally dependent innovations like language initiated a cognitive and behavioral revolution in our species. I suggest in this book that a set of conceptual tools called "numbers"—words and other symbols for specific quantities—is a key set of linguistically based innovations that has distinguished our species in ways that have been underappreciated. Numbers are, we shall see, human creations that, like cooking, stone tools, and the wheel, transformed the environments in which we live and evolve. While anthropologists and others have long been enamored with highlighting such inventions and their role in changing the script of the human story, the role of numbers has received insufficient attention in the past. The motivation for that inattention is simple: we are only now beginning to appreciate the extent to which the tools called "numbers" have reshaped the human experience.

PART 1

Numbers Pervade the Human Experience

1

NUMBERS WOVEN INTO OUR PRESENT

How old are you? From an early age, the answer to that question is literally at your fingertips. And it probably took only a fraction of a second for you to come up with an answer. Could there be an easier question, really? Many facets of your life are filtered by the number of your years. Can you drive a car by yourself? Well, it depends on how many years you have lived. Are you pleased with what you see in the mirror? That likely is influenced at least somewhat by your age, and what you expect to see in the mirror. Should you have a more fulfilling occupation? Hard to answer without knowing your age. The response to these and many other questions, which strike at the core of your identity and your day-to-day experience, can really only be stated if you know the answer to that first simple question. That question is undeniably meaningful to people in our own cultural matrix.

Yet, remarkably to those of us who attribute so much significance to our age, that same question is meaningless to members of some other cultures. This is not simply because members of such cultures fail to keep track of the earth's revolutions around the sun, but because they do not have the means of precisely quantifying such revolutions. In other words, they do not have numbers. Among the Amazonian indigenes known as the Munduruku, for instance, there are no precise words for numbers beyond 'two.' In the case of their Amazonian counterparts the Pirahã, no number words of any sort are used, not even for 'one.' How then could the "how old are you" question be answered by speakers of these languages? Or what of other number-based questions that, to most of the world's people, also get at basic aspects of life? Consider a few more examples: What is your salary? How tall are you? How much do you weigh? In a world without numbers, such questions are useless— unaskable and unanswerable. These questions and their potential responses cannot be formulated, at least not with any precision, in anumeric cultures. And for much of the history of our species, all human cultures were anumeric. Numbers, the verbal and symbolic representations of quantities, radically transformed the human condition. In this book I explore the extent of that transformation, which has been remarkably recent. I focus on the transformative power of verbal numbers but also examine the role of written numbers. For terminological clarity, I usually refer to verbal numbers simply as *numbers,* and reserve the term *numerals* for written numbers. When referring to the abstract quantities described by numbers, I use symbols like 1, 2, 3, 4, and so forth.

During the past decade, a flurry of research has been done on numbers and numerals by archaeologists, linguists, psychologists, and others. From that research, a new story of numbers is beginning to take shape, a story that is told in this book. In short, it goes something like this: Despite what we once thought, numbers are not concepts that come to people naturally and natively. While quantities and sets of items may exist independently, apart from our mental experience, numbers are a creation of the human mind, a cognitive invention that has altered forever how we see and distinguish quantities. This notion is perhaps unintuitive to many of us who have lived our entire lives with numbers, having them coaxed into our mental experience from infancy. Yet, like another key interrelated symbolic innovation of our species—language—numbers are in fact a culturally variable creation. Unlike language, however, numbers are absent in some of the world's populations. They are an innovation that indelibly impacts how most, but not all, people construe much of their daily experience. This indelible impact is at the core of the story this book tells. We will examine the way that numbers, one of the key inventions in the course of our species' history, served as a sort of flint stone that ignited the human timeline.

This story involves a lot of pieces, and later in this chapter I outline the way this book attempts to step from one piece to another, along a coherent path toward a newly formed conclusion. But before we talk about those pieces, I should exemplify what I mean when I say that numbers transformed the human experience. Perhaps the best way to do that is to examine further how we perceive the passing of time. I have noted that, without numbers, you obviously cannot label the quantity of the earth's trips around the sun since your birth. But maybe, you might counter, you could still have some sense of how old you are. You could know you were born before your sister and after your brother, for example, so you could know you are older than the former and younger than the latter. And you could recognize the changes of seasons and appreciate that you have lived through previous seasonal cycles. So you could at least know you are many years old, and perhaps know that you have experienced comparatively more or fewer years than your contemporaries. Yet, as we will see in our discussion of anumeric peoples in Chapter 5, such a sense of age is vague if one does not have recourse to numbers. The role of numbers in our temporal perception is more apparent, though, when we consider the passing of time at its most basic level—apart from how we enumerate years.

This consideration requires a brief digression into our general understanding of time. In some ways, time is a difficult notion to grasp as it is inherently abstract. What does it mean to perceive or feel time? Well, it turns out, it depends on whom you ask and what culture they are from, or which language they speak. Recent research has demonstrated that time is conceived of in disparate ways across some populations. Next, I address some of this cultural variation, and then I will suggest that numbers have played an ineffable role in shaping the culturally variable experience of time.

We often talk about the 'passing' of time or of time 'passing by.' In fact I have done so in preceding paragraphs, and I doubt you considered such phrasing unusual. We also speak of time moving 'slowly' or 'quickly,' but clearly all these manners of speaking are metaphorical. Time does not really move, nor do we move through it. Cognitive scientists established some time ago that humans have a pervasive tendency to utilize concrete things, such as objects moving spatially, to metaphorically describe abstract aspects of our lives, like time. So we can talk about the 'movement' of time, or conversely about 'going through' a tough time, or of 'seeing' a difficult time 'ahead,' or of our inability to go 'back' to the past, or of choosing the right career 'path,' or of facing a fork in the 'road' of our lives, and so on. For speakers of English and many other languages, there are countless expressions that reflect and reify spatial interpretations of time. And most prominent among these metaphorical orientations is the one that pervades the examples just given, in which we face the future as time passes through us. It turns out, though, that for speakers of some languages, time does not work this way. For speakers of Aymara and several other languages, the future does not lie in front of the speaker. In fact, for the Aymara the future lies behind the speaker, while the past is located metaphorically in front of the speaker. This orientation is evident in various expressions about time and in the hand gestures fluent Aymara speakers make when talking about past and future events. (Arguably, such a metaphorical orientation maps more directly onto the human experience since we can already 'see' what has happened in our past.) So, some humans perceive the 'movement' of time in a manner that seems diametrically opposed to the way we describe it and perceive it.[1]

The malleable spatial basis of temporal thought is further evident when we consider another way in which we can metaphorically depict time, namely, as moving from left to right along a measurable line. In our culture and others, there are myriad ways in which time is depicted as such. These include calendars, the progress bars on Netflix and YouTube, timelines in history books, and so forth. And robust experimental evidence suggests that such default symbolic practices impact how we perceive time. For example, when Americans are given a set of pictures depicting events at different stages (for instance, pictures of a banana being peeled and eaten) and are asked to orient those pictures from first to last, they typically place them in a left-to-right order so that the earlier images are closer to the left side of their own body. When members of some other cultures are given the same task, however, the ordering changes. Recently linguist Alice Gaby and psychologist Lera Boroditsky found that, in the Thaayorre culture on the Cape York Peninsula, people order the pictures not from left to right, nor from right to left (a pattern that surfaces in some cultures). Instead, they orient the pictures according to the trajectory of the sun, with earlier images being placed toward the east and later ones toward the west, regardless of the direction the person organizing the images is facing.[2]

Such findings reflect an important point: How we think about time is largely a matter of cultural and linguistic practice. And here is where numbers come into the story of how we make sense of this fundamental facet of our lives, because numbers clearly impinge on how we think about the 'movement' of time. Whether we think of time as passing through us or as moving along a timeline in front of us, its 'movement' is divisible and countable. Think again of progress bars in online videos and how numbers (denoting minutes and seconds) track the icon that represents the moment being displayed in the video. In fact, numbers are ubiquitous in spatial, symbolic representations of time like left-to-right calendars and timelines. This number-centric conceptualization of time arguably governs our lives.

What time is it? For me, as I write these words, it is 10:46 A.M. on the east coast of the United States. Since it is that time of day, I am in my office, at my desk, rather than at home or some other place. But what does that time really mean? Well, it means it has been ten hours and forty-six minutes since midnight, sure, but that is a tautological restatement. What are hours? What are minutes? In truth, they do not exist apart from our mental and numerical experience. They are simply an arbitrary means of quantifying our existence, of dividing the metaphorical passing of time into discrete units. They are an indication of the fact that humans at some point chose to quantify time, to number moments of experience. Time may be real, existing apart from our own experience, but hours and minutes and seconds exist only in our minds, as a way to engage with the world. This means of engagement is itself due to particular linguistic and cultural traditions. Such time units as hours, minutes, and seconds are actually the detritus of ancient number systems. These units are really just linguistic vestiges of extinct civilizations.

Consider the division of each of the earth's rotations, each day, into 24 hours. Why is each day so divided? There is no astronomical motivation for this division, after all, and we could in theory have any random number of hours per day. But our time-keeping system owes its existence in large part to a tradition begun by the ancient Egyptians, who developed sundials more than 3,000 years ago. Those sundials were designed to partition the daylight into twelve equal portions. This twelve-fold division was simply a by-product of the Egyptians' choice to divide the daylight in a culturally appropriate manner, as measured by shade along sundials. The choice allowed for ten divisible units of sunlight from sunrise to sunset, a natural choice since ancient Egyptian had a decimal number system like our own. Yet the sundial creators also added a unit each for dawn and twilight, the periods of the day that were not dark but in which the sun was not visible over the horizon. The simple decision of the Egyptians to divide daylight in such a manner yielded units of time based around the number 12, giving days a duodecimal feel. As we will see in Chapter 3, there are many bases in the world's spoken number systems, and duodecimal systems are pretty uncommon (and somewhat confusing to many people familiar with, say, decimal systems). Yet, because of the choice made by ancient Egyptian timekeepers, our language and thought about time are based in large part on a duodecimal-like system. This system is now firmly ingrained in our lives and enforces a certain perspective on our days. The existence of twelve-hour nights is also due to the Egyptians, as is, more indirectly, the 24-hour day / night cycle so familiar to us all. The latter system was more formally codified by Greek astronomers in the Hellenistic period, though hours of an exact and equal duration could not be appealed to until precise time-keeping mechanisms were invented. (The pendulum clock, a key innovation in time-keeping, was not created until the mid-seventeenth century.) Ultimately, then, the existence of hours is a historical accident. Had the Egyptian sundials originally separated daylight into tenths instead of twelfths, we would have ten major time units per day and night, respectively. The earth's rotations would be divided into twenty 'hours.'[3] In fact, a decimal-based time-keeping system was implemented in France immediately following the revolution, but the system failed to catch on due to the cultural entrenchment of hours and minutes. It is apparently easier for a nation to dethrone a monarchy and decapitate a sizeable portion of its citizenry than to reorient itself to new time units.

Minutes and seconds are also the result of culturally and linguistically contingent decisions made long ago. These units of time owe themselves to the sexagesimal (base-60) system employed by Babylonians and, before them, by Sumerians. These cultures appear to have been the first to use such a base for astronomical calculations, for reasons that remain nebulous. Some believe the sexagesimal system gained prominence in Mesopotamia because it is neatly divisible by 1–6, as well as by 10, 12, 15, 20, and 30. Others think such base-60 systems likely arose because humans have five digits on one hand to count the twelve joints on the nonthumb fingers of the other hand (and $5 \times 12 = 60$). Regardless, sexagesimal systems are not common. They have only developed a few times during the history of the world's languages. Yet the sexagesimal nature of the Babylonian counting system is the reason minutes and seconds last as long as they do—because those are the units of time you arrive at if you divide hours and minutes, respectively, by sixty. People can now rely on independent metrics to define seconds, for instance the duration of a predefined number of energy fluctuations in a cesium atom. This definition serves as the standard of the atomic clock. But such a metric was chosen only because it closely approximated the length of traditional seconds that were merely a by-product of an ancient number system that yielded an effective but arguably unwieldy means of referring to time.

In sum, our construal of time is impacted by the metaphorical mapping of time onto space. Crucially, though, that space-based view of time is quantified in ways that are completely dependent on the existence of numbers. More specifically, this quantification is dependent on the characteristics of number systems once used in places like ancient Babylon. How we think about time—in discrete quantifiable units of hours, minutes, and seconds—is due to the features of extinct languages and cultures, features with vestiges in our contemporary lives. These vestiges continuously orient how we organize our everyday experience. So ancient numbers with eccentric characteristics continue to shape the way we experience time, this abstract yet fundamental part of life. Our lives are, after all, governed by hours, minutes, and seconds. Yet time does not actually occur in these or any other discrete units. The segmentation of time into quantifiable units is truly a figment of the human mind.[4]

This discussion of the role of numbers in shaping our perception of time is illustrative of how powerfully numbers, and differences among number systems, can impact our cognitive and behavioral lives. Yet we will see during the course of this book that the invention of numbers impacted our lives, and the human narrative more generally, in many other equally profound ways. Before talking about those ways, though, some relevant background on our species is in order. This background is essential to, and intimately linked with, the story of numbers that this book tells.

Young *Homo sapiens*

Our capacity for measuring the passing of time is quite handy when discussing the recent origins of *Homo sapiens*. Numbers help depict just how young our species is: The universe is about 13.7 billion years old, the earth about 4.5 billion, and eukaryotic life about 3 billion. The emergence of primates occurred sometime around 65 million years ago. The fossil record suggests that hominins, including the ancestors of humans, have lived for only about a tenth of that time. Much debate remains about when exactly we, modern humans, first emerged, but we have definitely been around at least 100,000 years. Accepting the latter figure for the moment, this means we have only existed about one year for every 130,000 the universe has been around. This is an often unrecognized feature of humans: we are really, really young. Yet, despite our youth, we have in many ways shaped this planet on which we have been residing for such an insignificant fraction of its existence, particularly in the past few thousand years. Numbers, we will see, are a big part of how and why that happened.[5]

Extensive data demonstrate that *Homo sapiens* and its ancestral species evolved in Africa. Key components of our present physical characteristics began to take shape there, for instance the bipedalism first clearly evident in australopithecines—whose footprints from 3.7 million years ago are apparent in the volcanic ash in Laetoli, Tanzania. Larger brains also emerged in species like *Homo erectus* (about 1.8 million years ago) and *Homo heidelbergensis* (more than half a million years ago), species that managed to explore non-African continents but whose material record is not as suggestive of a dramatic cognitive leap forward, as in the case of *Homo sapiens*. This latter point hints at something crucial: human ancestors had relatively large brains, though not as large as our own, long before we arrived on the scene. Yet, despite their large brains, the behavior of our closest ancestral species was not nearly as remarkable when contrasted to other great apes. It bore little resemblance to that of modern humans or of *Homo neanderthalensis,* our sister species that lived in Europe for about half a million years—until its extinction was apparently accelerated by our arrival on that continent.[6]

So one reasonable way to frame the evolution of our species is as one of radical recent change. Sure, our lineage has been evolving for millions of years in ways that made us, physiologically, who we are today. Yet for most of that time our ancestors lived harsh, short lives, often serving as prey for larger African species. We did not always outcompete other species to the degree that we do now. I recently spoke with a fellow anthropologist, a paleoarchaeologist who studies the fossils of various hominin species in Africa. He mentioned that one of the most striking features of these fossils is the violence they reflect. Many are characterized by osseous lesions and fractures, and they often bear the dental imprints of predators and scavengers. Frequently the fossils are located in the lairs of predators such as lions. Most fossils are of children and young adults. Such evidence bleakly suggests that many of our ancestors lived brief and difficult lives, struggling to compete with neighboring predators.

Much of this struggle could be said to be the result of an apparent cognitive stagnancy. This stagnancy is evident in the gradual material innovations apparent in the fossil record for several million years. Consider the stone hand axe, referred to by anthropologists as the Acheulean axe, first developed by *Homo habilis* around 1.75 million years ago. This hand axe, portable and eminently practical, was a crucial tool for our ancestors. Yet it was a staggeringly simple tool contrasted with, say, the atlatl or the bow and arrow. And somehow hominins relied almost exclusively on it for more than 1.5 million years. With their bipedalism, relatively large brains, and simple tools, our ancestral species appear to have been on the launch pad to modernity for hundreds of millennia. Yet the launch failed until a recent ignition.

After our ancestors' fight for survival during most of the Paleolithic era, things took a sharp turn for the better. (The Paleolithic era lasted from about 2.5 million years ago until about 10,000 years ago.) At some point in the past 200,000 years, likely around 100,000 years ago judging from the archaeological record, there was a seemingly radical shift in how our ancestors thought. This cognitive shift is evident, for instance, in the intricate and polished bone tools uncovered in the Blombos Cave in South Africa, along with other artifacts in that cave and in others, discussed in more detail in Chapter 10. Shortly after such tools were invented, humans began leaving Africa in earnest. Genetic analyses of humans alive today suggest that modern non-African peoples are the descendants of a small set of *Homo sapiens,* whose African exodus plausibly took them through the Red Sea at the straits of Bab al-Mandab.[7]

What happened next was as unprecedented as it was unpredictable, given that humans were a struggling species that faced the very real threat of extinction. While other primates left Africa accidentally, primarily to end up in other tropical biota, our ancestors began a process of willful exploration that persists today. Through a global circumambulation that lasted dozens of millennia— until humans reached the tip of South America some 14,000 years ago—we eventually adapted to nearly every global environment. We outcompeted species in harsh environments like the Siberian tundra, the Tasmanian bush, the Atacama Desert, and nearly every biosphere in between. The archaeological record speaks to our advancement. Quite simply, humans became adapted to adaptation. This newfound adaptation would, of course, have been impossible without language and culture, the most distinguishing characteristics of our species.[8]

The origins of language and culture are still a matter of great debate. According to the work of many anthropologists, the human linguistic and cultural revolution was due in large measure to a greater reliance on collaboration. This reliance was two-fold: first, humans were forced to rely on collaboration to outcompete other species, and second, particular groups of humans relied on more advanced forms of collaboration when competing with other groups of humans. This account is supported by the fact that humans, while not apparently hardwired genetically for language specifically, are predisposed to collaborate intentionally with other members of their species. Human infants, who lack some of the higher cognitive functions of other great apes, are keen perceivers of potential collaborations with other individuals. The installation of collaboration into the human hardware appears to have been, at the least, an important precursor of a shift from the basic gestural communication systems of apes to the more robust speech-based communication system of humans. In other words, what makes us linguistic creatures is not so much that we are provided innately with a specific set of linguistic skills but that we are able to cooperate and collectivize our cognitive skills, many of which are evident in other more disconnected apes. This move toward cooperation appears to have played a pivotal role in our cognitive lives, yielding a communicative shift that helped make us distinctly human. Language would have been impossible without our emphasis on cooperation, along with the associated attention we began paying to the ideas and intentions of others. Whatever its origins, it is indisputable that language refashioned the human experience and allowed us to excel before and after we left Africa.[9]

Language shapes how we think, even facilitating certain kinds of nonlinguistic thought. More advantageously, language allows new forms of cooperation and enables humans to transmit the solutions to ecological challenges, both within and across generations. Words, the conduits of ideas, are cognitive tools that allow people to record and convey solutions to the panoply of novel problems they face as they enter new environments. The innovation of language allowed humans to access the ideas in the minds of other humans and transmit those ideas effortlessly, without having to continually generate novel ones. It enabled the cross-generational cultural ratchet I mentioned in the Prologue. We remain well adapted to our current environments, even in urban settings in a modernized world, because ideas have been transmitted to us, though language, from the minds of others since our infancy. Language and other symbolic cultural practices allow us to store and access concepts easily, including basic concepts that enable our individual and cultural survival.[10]

While a definitive account of the emergence of language is lost to time or stuck tantalizingly in the penumbra of the archaeological record, its significance as suggested here is uncontroversial. It is clear that words and other symbolic representations serve(d) as incisive tools, likely the greatest set of tools we ever acquired. Yet there is a subset of this set of verbal implements, the cognitive tools of numbers, that played a particularly trenchant role in sculpting humanity since its African exodus and likely before we even left Africa. This subset of verbal tools enabled us to see and manipulate quantities in new ways. As already discussed, the specific tools in question enabled us to perceive time in new ways as well. These numeric tools, this book suggests, also led to the advent of agriculture and writing, and indirectly to the technologies that flowed from the latter two phenomena. They are tools that forever altered our conceptual and behavioral experience.

Quantities in Nature, Numbers in Our Minds

Often the function of words is to label preexisting objects or ideas. For instance, the word 'panda' labels a certain species of mammal. That species exists regardless of the existence of the label. But sometimes words denote concepts that do not really exist apart from the existence of the words in question. Consider the case of color. We constantly interact with the visible portion of the light spectrum, a minor segment of the range of electromagnetic waves. This visible light spectrum is continuous, with no definitive physical divisions. So, for example, there is no actual point on the light spectrum at which blue and green are neatly split. Instead, green and blue blend into each other. It is for this reason that many languages dispense with terms such as 'green' and 'blue,' employing instead a word for the color category 'grue.' However, speakers of languages like English refer to this color contrast all the time. In doing so, they essentially bring a clearer distinction between 'green' and 'blue' into existence. They use words to communicate about portions of the light spectrum that can be roughly discriminated but that do not have absolute boundaries. Speakers of some other languages divide the color spectrum up in different ways. For instance, the Berinmo of New Guinea use the terms 'wol' and 'nor,' words that distinguish portions of the light spectrum referred to by the same English term, 'green.' Such crosslinguistic differences yield subtle but reliable effects on the way speakers of such languages perceive and recall colors. In short, color terms do not simply label preexisting color concepts shared by all humans, they also call into existence more rigidly bounded color concepts.[11]

Much as color terms help us demarcate and concretize certain portions of the light spectrum, words and other symbols for numbers generate specific kinds of quantities in our mental lives. It turns out that humans do not 'see' divisions between most quantities without numbers. In the absence of numbers, the way we see quantities of objects in our natural environments would not be that different from many other species. Were it not for our capacity to innovate and adopt numbers, we would not have the tools that are prerequisite to navigating, purposefully and with direction, the sea of quantities around us.

It may seem odd to suggest that numbers are a human invention. After all, some might say, regardless of whether humans ever existed, there would still be predictable numbers in nature, be it eight (octopus legs), four (seasons), twenty-nine (days in a lunar cycle), and so on. Strictly speaking, however, these are simply regularly occurring *quantities.* Quantities and correspondences between quantities might be said to exist apart from the human mental experience. Octopus legs would occur in regular groups even if we were unable to perceive that regularity. *Numbers,* though, are the words and other symbolic representations we use to differentiate quantities.[12] Much as color terms create clearer mental boundaries between colors along adjacent portions of the visible light spectrum, numbers create conceptual boundaries between quantities. Those boundaries may reflect a real division between quantities in the physical world, but these divisions are generally inaccessible to the human mind without numbers.

Number words representing quantities have often been considered convenient labels for concepts that humans are innately endowed with or naturally come to learn during biological development. In contrast, more recent work suggests that numbers are not simply labels. As linguist and number specialist Heike Wiese has insightfully noted, "language gives us instances of numbers, words that we can employ as numbers, rather than just as names that we employ to denote numbers and to reason about them."[13] Most specific quantities do not exist in our minds in the absence of numbers. This claim may surprise some, but it is a claim that is well grounded empirically. In contrast, the assumption that numbers are merely labels for preexisting ideas is not actually well supported. It turns out that humans, like other animals, cannot precisely and consistently grasp exact quantities beyond three unless they have numbers. Beyond three, we can only estimate the quantity of objects we are perceiving if we do not know numbers. This finding has been supported with recent experimental work, conducted by many scholars (including me), with number-less people. It has also been supported by research with infants and other prenumeric children. Such results will be discussed in detail in Part 2. As we will see, our innate obstacles to distinguishing quantities can only be dismantled by the tools of numbers.

Admittedly, though, this account raises a paradox: If humans cannot think exactly about quantities without numbers, how did they ever come up with numbers in the first place? The first point to make in response is that, at least in some respects, this paradox applies to any human invention. For any invention to happen, humans must first recognize a concept that they do not typically and naturally recognize. Inventions are not genetically hardwired but are made through a series of often simple realizations. We are not innately predisposed to think of things like fulcrums, screws, wheels, hammers, or other basic mechanical tools. Yet through a variety of realizations, each of these tools was developed. Consider the wheel, such a simple and practical tool. It is almost hard to imagine that humans could *not* invent this tool, given our awareness of rolling round things in our natural environment. Yet the wheel, along with the axle, is a fairly recent innovation that most past cultures lived without (including some large societies like the Inca). So, despite its simplicity and the ease with which it is grasped conceptually, humans do not have an inborn 'wheel' concept. Similarly, a verbal tool such as the word 'seven' seems remarkably intuitive once we are presented with it. Yet some people are not familiar with the exact quantity it denotes. Much as they may not be familiar with wheels but quickly understand their usefulness when presented with an actual wheel, they learn to grasp the concept of exactly seven things only once they learn the word that instantiates that concept. For that simple reason, number words facilitate not just complicated math but also the mere differentiation and recognition of quantities greater than three. (Experimental support for this conclusion is discussed in Part 2.)

But, as you might have noticed, I did not completely resolve the paradox. Phrasing it in another manner, we might ask: How exactly did individuals without numbers ever come to appreciate how such words could represent quantities, if numbers are crucial to the recognition of precise quantities? As a promissory note to a fuller account laid out in Part 3 of this book, consider this: Some members of our species have clearly realized at different points in time that an existing word can have its meaning extended to represent a specific quantity beyond three. (They recognized, for example, that 'hand' could refer to 5, not just to a physical appendage.) It is this simple realization that is at the heart of the invention of numbers. Yet this is not a realization we are born with as a species, just like we are not born with the realization that wheels can exist, or that steel ships can float, or that aluminum planes can fly. But when some number-inventors happened upon the realization that words could be used to distinguish quantities like five from six, it enabled them to establish a new way of thinking about quantities that others began to adopt. Through that adoption, numbers spread.

As I suggest in more detail in Chapter 8, the fact that some humans have been able to invent numbers is largely the result of anatomical factors. The simple realization that precise large quantities exist and can be labeled has usually been due to the fact that we have regularly occurring quantities right in front of our faces. We have five digits on each hand. Our biology constantly presents us with matching sets of five items that we are not cognitively preordained to recognize, just like other species are not. Yet humans have been able to occasionally recognize this correspondence. This is a seemingly straightforward realization, but the mere recognition of such a biological correspondence does not necessarily yield numbers. Quantities, even the five digits on each hand, can potentially be recognized in only a fleeting manner. However, when words like 'five' are introduced and are then used productively to describe the quantity of digits on each hand, numbers are invented. This common anatomical route toward the invention of numbers is supported by much linguistic data, such as the frequent similarity between the word for 'five' and the word for 'hand' in the world's languages. (This point is made in detail in Chapter 3.)

The invention of numbers, made at various times during the course of human history, did not simply facilitate our thinking about quantities. Numbers enabled the precise and consistent discrimination of quantities greater than three. This hypothesis will be fleshed out more fully during the course of this book. For now, though, I have hopefully imparted a clearer sense of what I mean when I say numbers are revolutionary and invented conceptual tools. This book suggests that the invention and widespread adoption of these tools ultimately resulted in a cognitive and behavioral reorientation of humanity. They were perhaps the single most influential tools in the linguistic toolkit that enabled the recent transformation of our species as discussed in the previous section. Furthermore, they enabled or at least facilitated all sorts of more recent innovations discussed later in this book. Without these practical cognitive tools, there would likely have been no agricultural revolution and certainly no industrial revolution.

Where This Book Will Take Us

This book presents a synthesis of anthropological, linguistic, and psychological evidence. It considers data from human populations as well as data from other animals. All of those data lead inexorably to the simple conclusion already presaged: numbers served as fundamental conceptual and behavioral scaffolding, helping establish the larger edifice of modernity.

In the remainder of Part 1, we examine how pervasive numbers are to the human experience, focusing on symbolic representations for quantities in the archaeological and written records (Chapter 2), as well as in speech. We survey number words (Chapter 3) and other linguistic references to quantities (Chapter 4) in languages around the world. The data presented in these chapters suggest that numbers serve as a key component in nearly every one of the world's languages as well as in ancient nonverbal symbolic systems. Furthermore, the findings surveyed underscore the importance of human anatomy and neurobiology in the creation and usage of numbers.

In Part 2, we will take a look at the role numbers have had on humanity by detailing relevant findings gathered with adults that are unfamiliar with numbers (Chapter 5). We will also examine the numerical cognition of prelinguistic children (Chapter 6) and the numerical capacities of other species, many closely related to our own (Chapter 7). This examination will focus on recent studies conducted by anthropologists and linguists in often remote settings, along with studies by lab-based researchers in other branches of the cognitive sciences.

In Part 3 of the book we will consider how numbers have shaped most contemporary cultures. We will take a look at how numbers and basic arithmetic were likely invented (Chapter 8). I also suggest that numeric language helped transform human patterns of subsistence (Chapter 9). We will see how numbers enabled a florescence of other material and behavioral technologies, technologies that led to key milestones in recent human history. Finally, the book concludes with a consideration of some pivotal ways in which numbers have, at least indirectly, changed human cultures both socially and spiritually (Chapter 10).

2

NUMBERS CARVED INTO OUR PAST

Perched high above the forest floor deep in the heart of the Brazilian Amazon, near the quaint town of Monte Alegre, is an array of paintings on hillside cave and outcropping walls. The paintings, created by indigenous artists more than 10,000 years ago and meticulously documented by archaeologist Anna Roosevelt, have helped alter our understanding of precolonial history in the Americas. On one of the tableaus is a group of painted 'x' marks in a gridlike arrangement. The functions of this particular painting, more chart than art, are uncertain. But the marks quite likely reference quantities—of days, full moons, or some other valuable cycle lost to time. The painting is indicative of a larger trend. Over the past several decades, archaeologists have discovered numerous pieces of evidence that ancient peoples were paying attention to quantities. And they were depicting these quantities in two dimensions. Not with full-fledged symbolic writing, but with painted marks on cave walls, as well as carved marks in wood and bone. Such tally marks are symbolic in the sense that they stand for something else. But they do not represent quantities in a fully symbolic, abstract manner like true numerals, for example the way the numeral 7 refers to a set of seven items regardless of the type of item in question. Such primeval tallies could be termed *prehistoric numerals*—quasi-symbolic precursors to modern written numerals. Consider that the Roman numeral for 3 is III, much like three items would be represented if one were simply tallying them. Even our own numerals of Indian origin have clear vestiges of a tally system. After all, the numeral 1 is depicted with just one line like a simple tally mark.[1]

About 5,000 kilometers away from Monte Alegre, at Little Salt Spring, Florida, archaeology students from the University of Miami recently discovered a remarkable portion of reindeer antler also dated to about ten millennia before the present. A recent photograph of the antler is reproduced in Figure 2.1. As is evident in the picture, there is a series of lines carved into the side of the antler. The lines are very regular and are about five millimeters long. Moreover, the spacing between each line is quite consistent, suggesting that the marks were made deliberately and systematically. Next to these marks are smaller etchings in one-to-one alignment with the larger incisions. These miniscule, secondary notches suggest that the bone was used to keep track of the progression of something, and that quantities were ticked off along that progression. (In Figure 2.1, the secondary etchings are slightly to the left of the main lines on the antler.) The significance of the antler segment has gone largely unnoticed, as it was only recently described in a niche anthropology journal, without reference to its larger implications. Unlike the case of the Monte Alegre painting, though, one can advance a highly plausible hypothesis as to the function of the antler's marks. In fact, the marks suggest that this piece of antler is the oldest known New World artifact used for calendrical purposes. Several pieces of evidence support this conclusion.[2]

The water in Little Salt Spring is anoxic (lacks dissolved oxygen) at depths greater than 5 meters below the surface. The piece in question, a cleanly cut segment of antler about 8 centimeters long and weighing about 50 grams, was found at a depth of 8 meters. It has been surrounded by anoxic water since it was cut about 10,000 years

2.1.   The reindeer antler of Little Salt Spring, Florida, with a colleague's hand for scale. Photograph by the author.

ago. Artifacts do not deteriorate in anoxic water like they do in regular water, so the antler has been meticulously preserved. We can be confident, therefore, that the number of marks on its side exactly match those engraved by some artisan all those years ago. Furthermore, the antler was found embedded in the ground next to the edge of an underwater cliff. This cliff was not submerged during the glacial period in which the artifact was made, when water levels around Florida were much lower than they are today. The sloping summit of the cliff served as a hunting site during that time. Numerous faunal remains and weapons have been uncovered there by University of Miami marine archaeologists John Gifford and Steve Koski and their students. This research team has carefully described and dated the remains and weapons in question, and has found that they belong to the same era as the carved antler. Given that the reindeer antler in question was uncovered at this site, it is reasonable to assume the bone was used for some purpose associated with hunting. The latter conclusion is supported by another crucial piece of evidence, namely, that twenty-nine major incisions were carved into the antler segment. There is now a removed chunk where one of those incisions used to exist, as evidenced by a smaller notch next to that removed chunk. One of the middle incisions is less regular, so it is possible that only twenty-eight of the incisions were made intentionally. However, the latter possibility seems doubtful, given the regular spacing between the marks, as evidenced in Figure 2.1.

Since Little Salt Spring was clearly used as a Paleolithic hunting site, it is likely that the antler's marks represent days or nights. Lunar phases impact hunting practices due to such factors as the altered behavior of some animals during full moons and the influence of moonlight on hunters' visual acuity. The antler's array of twenty-nine incisions, then, likely depicts the number of days in a lunar month. A synodic month lasts, on average, 29.5 days. This calendrical interpretation is also supported by a subtler piece of evidence on the antler: there is no smaller notch next to one of the endpoint lines in the tally of marks (the bottom endpoint line in Figure 2.1). This suggests a smaller notch was unnecessary on the last instance of whatever had been tallied by the larger adjacent mark. In other words, checking off the last mark was apparently redundant. This would certainly be the case if a hunter were keeping track of the lunar cycle, since the occurrence of a full / new moon would not need to be checked off on the actual night in question. The hunter would already be well aware that the full or new moon had arrived. Given such factors, along with the hunting-associated locale in which this artifact was found, the likely and noteworthy implication of the antler piece is that hunters used it as a tool for counting and recounting the days / nights of the month. In other words, more than 10,000 years ago and not far from present-day Miami, people were using linear marks to keep track of quantities. These prehistoric numerals were tally marks made in a portion of reindeer antler that was cut to a convenient size, comfortably held in the palm of one's hand. Essentially, this was a Paleolithic pocket calendar, fortuitously preserved in anoxic waters. (Presumably many others like it have not been preserved.)

While the antler 'calendar' at Little Salt Spring may represent one of the clearest cases of a Paleolithic tool used for tracking the lunar cycle, the people who used it were certainly not the only Paleolithic humans to use tally marks on bones to keep track of quantities. At Grotte du Taï in southern France, for example, a small engraved bone plaque, also dating to the late Paleolithic, was discovered. This rib bone's surface has hundreds of engraved lines on it, and some analysis has suggested that the marks served a calendrical function. Another ancient artifact uncovered in France is the Abri Blanchard plaque, a 28,000-year-old bone with circular and oval engravings that likely represent the phases and movement of the moon. Furthermore, late stone-age people in Europe were apparently utilizing less intricate tally systems to represent quantities, much like the Upper Paleolithic peoples that lived in present-day Florida. This conclusion is supported by an exemplar of a simple tally system also discovered in France, the Abri Cellier avian bone that is of similar age as the Abri Blanchard artifact. The Abri Cellier bone contains linear marks that are fairly regularly spaced, not unlike the Little Salt Spring antler. The marks do not have adjacent, smaller notches like those evident on the Little Salt Spring piece, and they do not represent 29 or some other quantity with decipherable motivations for tallying. Still, recent analysis suggests that the Abri Cellier bone, like the Abri Blanchard and Grotte du Taï artifacts, offers evidence its creator(s) were deliberately representing numerical concepts materially.[3]

So it appears that humans in Europe, South America, and North America have been representing quantities two dimensionally for many thousands of years. We do not know for sure whether these prehistoric numerals were used along with number words. But given the role that number words play in facilitating mathematical thought and in the recognition of recurring quantities (see Chapter 5), these artifacts do hint at the usage of numerical language by their creators. It is unclear how long humans have been using such engraved and painted prehistoric numerals, but it is likely many dozens of millennia. In this chapter and Chapters 3 and 4, I highlight a simple finding evident in global archaeological and linguistic data: humans have long been occupied with the representation of quantities. Terms for quantities play a pervasive and nearly universal role in contemporary human languages, suggesting their prominent role in the history of the spoken word. Similarly, the numerical focus of humans is prominent in the archaeological record and in the history of writing systems. Numbers are quite literally carved into our historical record.

As in all discussions of the evolution of human symbolic systems, our focus inevitably shifts back to Africa. More specifically, our attention is drawn to a small Congolese region where, in 1960, Belgian geologist Jean de Heinzelin discovered an engraved baboon fibula about 15 centimeters in length. Subsequent dating demonstrated that this bone, named the Ishango bone after the eponymous place on Lake Edward where it was found, is at least twenty millennia old. Along the sides of the roughly cylindrical bone are three columns of etchings with marks clearly grouped into sets. Since the bone's discovery, there has been intense debate regarding the significance of these sets. Some have suggested that the groupings indicate the usage of a duodecimal (base-12) number system, or an awareness of prime numbers, or of a decimal system. Hypotheses are bountiful because, in truth, we do not know exactly what specific purpose the bone served. Here is what we do know: the notches along the sides of the bone are approximately parallel to the others in their respective columns. (The notches do vary slightly in orientation, and they also vary a bit in length.) More significantly, the quantities evident in the groupings of marks are not random. The first column contains the following number of notches, in order from top to bottom: 3, 6, 4, 8, 10, 5, 5, 7 (total = 48). The second contains sets of 11, 21, 19, and 9 (total = 60), respectively. Like the second, the third row also contains 60 notches, but in sets of 11, 13, 17, and 19. This last column contains only prime numbers, most likely coincidentally. What seems less coincidental, however, is that each of the last two columns contains the same total number of marks: sixty. Also, there is a distinct possibility that the first column reflects some doubling pattern, given the presence of adjacent groups of 3/6, 4/8, and 5/10, respectively.[4]

Perhaps due to the various tantalizing hypotheses regarding the marks along its sides, a simple but crucial fact about the Ishango bone is occasionally overlooked: one end of the bone has a sharp piece of quartz sticking out from it, an affixed point that was apparently used for engraving. It seems the Ishango bone served as a sort of stone-age pencil. Someone once held the bone between their fingers to make marks on other objects, likely on other bones. The significant implication here is that the sides of the bone may have served as some kind of numeric reference table for the person holding it, as they in turn detailed the quantity of items or events into the side of some other bone or perhaps a piece of wood. In other words, the bone served a real, yet abstract, purpose. Like some Paleolithic slide rule, it has quantities represented on its sides, perhaps to facilitate the accurate reproduction of those and other quantities. The bone suggests that some African populations were producing and reproducing prehistoric numerals at least 20,000 years ago.

Other African bones with etchings on their side date back even further. The same could be said of some bones in Europe, such as a 33,000-year-old wolf bone, with fifty-five marks on its side, found in the eastern Czech Republic. Yet the function of most very ancient carvings is likely lost forever. Nevertheless, one African bone much older than the Ishango bone, dated via radiocarbon methods to 44,000–43,000 years before the present, did apparently serve a mathematical function. Found in the Lebombo mountain range that straddles the border between South Africa and Swaziland, this bone also has engraved lines on its sides. The Lebombo bone is a baboon fibula, similar in size to the Ishango bone but apparently used for a less sophisticated or, at least, more transparent purpose. The bone has twenty-nine lines carved into its side so, like the reindeer antler of Little Salt Spring, it quite possibly served to track the lunar cycle. While such an interpretation is not conclusive, particularly given that the bone is broken on both ends and not cut and marked as neatly as the antler from Little Salt Spring, the interpretation is nevertheless plausible in the light of the aforementioned centrality of the lunar cycle to human populations, and in the light of the fact that similar tally calendars are utilized by some contemporary African populations.[5]

What is clear from all these bones embedded in the archaeological record is that humans have been recording quantities materially, using prehistoric numerals, for tens of thousands of years. We have long been concerned with storing and tracking quantities, whether the 29 days of the lunar cycle or other naturally occurring sets. This long-held fixation is evidenced globally, having arisen in populations that came to reside in Florida, Amazonia, southern France, central and southern Africa, and no doubt many other locations with still-buried artifacts.

Few human technologies have had the sort of reign of influence that basic tally systems have had. More intricate tally mark systems came to play a prominent role in Europe and elsewhere during the past millennium, and some simple kinds of tally systems are still utilized today. As one of many potential examples of such contemporary practices, let us take a look at the tally system of the Jarawara. The Jarawara are a group of about one hundred indigenes living under the dense canopy of southwestern Amazonia, subsisting primarily by hunting and gathering. The people are still adept at their traditional survival methods, and many are also somewhat familiar with urban Brazilian life. As recently as 5 years ago, it was thought that these people lacked native numbers of any sort. As we will see in Chapter 3, though, it turns out that the Jarawara have for centuries been utilizing a verbal number system. Furthermore, they have also been using a material tally system. This tally system was carved not into bone but into wood. One example of the tally system is depicted in Figure 2.2, a picture of a small stripped tree branch on which a Jarawara man adroitly engraved a series of notches. The triangular notches on the branch are grouped in a regular fashion. They occur in groups of 1, 2, 3, 4, 5, and 10 separate engravings. The artisan who carved the etchings described to me their traditional usage: when referring to quantities, for instance the number of days a Jarawara man expected to be gone traveling, the man would point at the appropriate series of notches. For example, if he thought he would be traveling one week, he could point to a group of five and a group of two. Such a representation of quantities is incredibly useful, though as we shall see in Chapter 5, some Amazonian cultures have no analogous representations—tactile, verbal, or visual—for quantities. They are number-less. The Jarawara, in contrast, traditionally relied on a portable tally mark system not terribly unlike the one used by Floridian hunters at Little Salt Spring 10,000 years ago. Since their system utilized wood and not bone, and bearing in mind the Amazonian jungle's natural consumption of most human artifacts, the Jarawara's tally system would not have naturally survived in the archaeological record for so long. But, given the presence of ancient tally-like paintings at Monte Alegre, people have probably been keeping track of quantities for thousands of years in Amazonia, too. Many intriguing techniques for depicting quantities visually, like that of the Jarawara, have no doubt been lost to time due to material degradation—not just in Amazonia but throughout the world.[6]

Hundreds of kilometers to the southwest of the Jarawara's few small villages, at the edge of Amazonia, a very different sort of engravings has recently been discovered, engravings that are consistent with the ancient usage of numbers by an unknown group of people. These engravings are found not in wood or bone, however, but in the ground. They are a series of enormous geoglyphs, large linear ditches 2–3 meters deep. Viewed from above, these glyphs represent regular geometric shapes, such as circles and quadrilaterals. Some are squares with perfectly regular sides as long as 250 meters. Mysteriously, some of these glyphs date back as far as 2,000 years. They were covered by dense jungles for centuries until deforestation allowed for their accidental discovery from small aircraft. While the story of the architects who made such glyphs remains veiled by antiquity, it is clear that those people relied on regular mathematical correspondences to create the glyphs.[7]

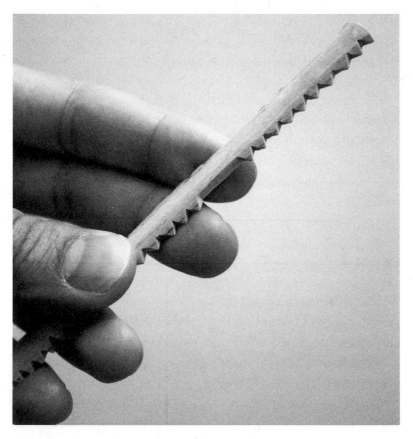

2.2.   The traditional tally system of the Jarawara. Photograph by the author.

Such geoglyphs were produced relatively recently when contrasted to some of the more explicitly mathematical prehistoric numerals we have mentioned. But the glyphs serve as a vivid illustration of a pervasive theme in archaeology: reflections of human numerical fixations are often evident in material records. This is true in the case of those most famous kind of Paleolithic remnants, cave paintings, as evidenced at the site near Monte

Alegre. While the functions of cave paintings are difficult to decipher, certain trends surface in the elaborate Paleolithic art on the inner walls of some of the world's caverns. And while the intended meaning of the paintings is impossible to discern definitively, the ages of such paintings can often be ascertained with confidence. Where the original artists used mineral paints like ochre to create their works, the tableaus are dated indirectly through the usage of organically based artifacts found near the art. Charcoal paintings, in contrast, allow direct radiocarbon dating of the paint used.

The combination of dating and interpretation reveals some extremely ancient motifs in European cave paintings. There is often an animal theme to the art. Aurochs and other bovines play a prominent role, as do bison, horses, and other large mammals. Yet another motif recurs in the cave paintings—that of the human hand. Tracings of the human hand are evident in the most ancient European paintings, such as those in El Castillo cave in Spain (about 40,000 years old), and in Chauvet (about 32,000 years old) and Lascaux (approximately 17,000 years old) in southern France. The hand shapes in these caves may have served some enumerating function, though this point is speculative. The hand stencils at the Cosquer and Gargas caves in France, dating to about 27,000 years ago, are more likely to have served some numerical function. The prints in these caves depict left hands with 1–5 extended digits. In all the depicted-hand configurations, the thumb is raised as though it represented the first number in a counting sequence. Archaeologist Karenleigh Overmann, who has conducted fascinating investigations of numerical representations in the human material record, has claimed that the relevant hand shapes in these caves indicate a technique of counting from the thumb (1) to the little finger (5). (That is, a solitary thumb represented 1 and, when all fingers including the little finger were raised, the quantity 5 was represented.) If one accepts such claims, they hint that other Paleolithic representations of the human hand could also have been used to represent quantities.[8]

What is particularly remarkable in the context of the present discussion is that the human hand and its fingers are a global motif in cave paintings, not just a European one. In fact, some of the oldest known paintings in the world, in the Sulawesi cave of Indonesia, are adorned with colorful hand tracings in which each finger is clearly visible. These paintings are about 40,000 years old. The Sulawesi paintings, like those in many other caves, were made when the artists blew a dye over a hand placed against a wall. Relatedly, some hand stencils in Australia's Fern Cave date back about 12,000 years. Even in South America this hand motif surfaces dramatically in the nearly 10,000-year-old art forms in the appropriately named Cueva de las Manos, the 'Cave of Hands' in the Patagonia region of Argentina. This cave contains a transfixing multicolored representation of dozens of hands, depicted in Figure 2.3.[9]

The representation of hands and fingers played a prominent role, evident across the world's continents, in the evolution of two-dimensional symbols and art. Judging from this global distribution of hand paintings, it is possible that humans practiced hand painting before they left Africa. The interpretation of such manual depictions is somewhat speculative, and in some cases the hands may have been represented simply out of convenience. Yet at least in some instances detailed analysis suggests a numerical function of hand stencils. Given the prominence of the hands in the linguistic representation of numbers, discussed at length in Chapter 3, and given the clearly numerical function of other ancient artifacts, such as the Ishango bone, it is not implausible that some of these artistic representations of hands also served basic quantitative functions. The hand stencils in the Cosquer and Gargas caves are particularly

2.3. Hand stencils at Cueva de las Manos, Argentina. Wikimedia Commons (CC BY-SA 3.0).

suggestive of counting. Such reasonable speculations aside, though, we must at least recognize that cave paintings are riddled with evidence that humans have long been focused on their hands—we are a manually fixated lot. As we shall see in our discussion of infant cognition in Chapter 6, the development of numerical thought in children is inextricably interwoven with this manual fixation. Even in utero we begin to pay attention to our hands. Our first attempts at representing quantities usually involve the usage of fingers. Furthermore, finger counting is a ubiquitous practice across the world's cultures.

Whatever the preponderance of hands in cave paintings may suggest about the history of numbers, it is evident that many creations of various Paleolithic artisans, made across geographic regions and over tens of millennia, depict quantities. Ancient carvings and paintings have a numeric interpretation in many cases. The tallying of quantities played a recurring role in the projection of human thoughts onto animal bones, onto wood, into the ground, and onto cave walls.

Why does the representation of quantities play such a prominent role in such ancient creations? The answer to this question is, I think, at least two-fold: first, quantities are easy to represent two dimensionally, compared to other basic aspects of the human experience like time (which is apparently represented indirectly through tallies of celestial cycles in Paleolithic artifacts) or emotions or particular physical locations, which require much more artistic sophistication to be accurately conveyed through drawing. In contrast, simple lines and other marks can easily and directly represent units or quantities. But this just begs the question as to why tallies so easily stand abstractly for other things, without literally depicting them. The answer here may lie, at least partially, in our hands—in the semblance of simple linear marks to our linelike fingers. In a way, we might say that fingers are anatomical, three-dimensional lines. It is not surprising, then, that members of many cultures around the world, though certainly not members of all cultures, have come to use lines to represent quantities much as they use fingers to do so. In other words, the transferal of finger-counting systems to tally systems requires a less dramatic cognitive leap than the innovation of other potential sorts of visual representations of concepts. Tally systems rely on a more straightforward projection of concepts onto two-dimensional space, practically speaking, when contrasted to entities and ideas that are trickier to represent artistically and that have no anatomical gateway to their simple representation.

Second and perhaps more crucially, carved numerical tallies—and, more hypothetically, painted finger-counting practices—became pervasive in the archaeological record because of their usefulness to those who created them. Representational means of encoding quantities are eminently functional. It is easy to see some potential advantages, for instance, of tabulating the precise number of men in a tribe one intends to raid, or alternatively of tracking the precise number of predators in one's vicinity. It is also easy to grasp some possible advantages of tracking the units of the moon's cycle. While humans can of course live and succeed without tracking such quantities, the advantages of doing so help explain why nearly all the world's cultures utilize numbers. Such advantages can improve ratios of survival, whether in warfare or hunting. The function of prehistoric numerals reached beyond the spiritual, social, or rhetorical. Prehistoric numerals could be key, at least in some instances, to mere survival.

It is no wonder, then, that prehistoric numerals played such a prominent role in the abstract representations made by humans during the Paleolithic. And their prominent role in such two-dimensional representations of ideas was not restricted to the stone age. It was also evident millennia later, as humans transitioned to more elaborate symbolic depictions of thoughts. For at the dawn of writing, the starring role of numbers once again played out on center stage.

Numbers at the Genesis of Writing

The Great Court of the British Museum in central London is sheltered by a steel and glass canopy. The translucent canopy functions like a sieve of light, somehow filtering out the gray London skies and casting an ethereal white glow on the impressive collection of artifacts housed underneath. The collection is peerless in many respects because of the manner in which the British empire gathered (or, more aptly in some cases at least, pillaged) finds from around the world. These finds include the famous Rosetta Stone, constantly surrounded by a paparazzi-like cloud of tourists as one turns left into the southwestern hall adjacent to the Great Court. But upstairs from the hall and comparatively ignored lies a smaller, plainer object that may yield greater insights into the development of human writing. Affixed to a wall in an unadorned manner, in a plain exhibit on the history of human writing, is a gypsum tablet that is some 5,300 years old (more than 3,000 years older than the Rosetta Stone). This tablet is only several centimeters long on each side and contains lines and dots made in the gypsum, all those years ago. These lines and dots, we now know, represent quantities—likely quantities of grain or some other item that featured in an economic transaction. They are more systematic than the marks evident in the Paleolithic record, as they are not just tallies of quantities. Instead they represent a standardized form of communication in two dimensions. They are the first true written symbols we know of, as each line and dot represents a specific abstract quantity. In other words, the marks in the gypsum are true numerals.

Out of the fog of scattered symbolic practices around the world, true writing emerged in Mesopotamia around the time this gypsum tablet was forged. The tablet now on display in London is an example of the transition scribes initiated in Mesopotamia, a transition from simple ways of depicting quantities to full-fledged writing. Epigraphers and other scholars typically distinguish writing, which fully encodes a particular language, from proto-writing, a more ancient form of symbolic practice (though less ancient than the prehistoric numerals used in Paleolithic contexts) that only depicts a limited array of possible meanings. The decision as to whether an ancient script should be considered proto-writing or actual writing is not easily arrived at and, in truth, such terms obscure the gradual process through which writing developed.

Terminological choices aside, it is generally agreed that writing was first comprehensively developed in the Fertile Crescent, more specifically in Mesopotamia, by the Sumerians. Yet it is not a uniquely Middle Eastern invention, as it was also developed independently in China and in Mesoamerica. From these regions, writing spread and evolved in accordance with local linguistic, social, and economic needs. Today there are dozens of writing systems, such as the one with which I am presently conveying my thoughts to you. Yet, in real and historically demonstrable ways, all these many writing systems can be traced back to one of three major writing traditions. Here our focus will be on the genesis of the oldest of true writing systems that developed in Mesopotamia. (In Chapter 9 we briefly discuss the origins of other writing systems.) Human writing was first born in this region, and the story of that birth reveals the essential role that numerals played in its inception.[10]

In a way, it is not particularly surprising that there has long existed a gravitational pull toward numeric symbols, since fluency with such symbols plays and has played an essential socioeconomic role in people's lives. Many of the best-preserved written symbols for quantities are coins or coinlike artifacts that encode specific monetary values. Given the limited rates of literacy for the majority of the existence of human writing, coins and other forms of currency associated with particular quantities have long been (and remain in some parts of the world) the only symbols that people were capable of interpreting. Even after its independent development in Eurasia and the Americas, true writing remained for millennia a niche skill practiced and passed down by a select group of practitioners in a few societies. Scribal classes are after all an economic luxury that resulted indirectly from agricultural practices that made such a specialized vocation possible. And the economic function of numbers played a distinct role in scribes' development of writing in agricultural Mesopotamia, a role that ultimately led to forms like the gypsum tablet above the Great Court. Let us briefly consider one theory as to how that happened.

Potentially as long ago as 8,000 years, people in Mesopotamia traded large quantities of agricultural produce and animals with one another. Ancient trade in the region was facilitated by the re- alization that quantities could be depicted symbolically and trans- mitted over long distances. One key method that developed in the region may seem archaic now, but it certainly must have been rev- olutionary at the time: solid clay containers were filled with tokens as a sort of contract. So, hypothetically, if one landowner agreed to pay another landowner a given amount of sheep, this agreement could be encoded in a clay ball. Tokens representing the specified quantity of sheep were embedded in the clay, which was then baked and hardened. The clay ball then served as a record that could be transported and later broken apart to confirm that a contract had been fulfilled. To facilitate this record-keeping, the actual number of tokens preserved inside the clay container could be denoted sym- bolically on the exterior of the container. With time, certain exte- rior symbols were also innovated to represent frequently recurring sorts of goods involved in the transactions. These exterior symbols could then be matched to the specified quantity described by the interior tokens. This matching system doubtlessly quickened the pace of economic transactions in a time predating actual mone- tary currency.

It is unclear how long this three-dimensional quantitatively oriented system for representing items persisted in Mesopo- tamia. Eventually, though, Sumerians began dispensing with the interior tokens altogether, and the system gradually transitioned from a three-dimensional one to a two-dimensional one. (Though the three-dimensional system may have persisted in some places.) That is, rather than encoding the quantities of trade items with actual tokens inside clay containers, the quantities were simply depicted on small clay tablets that came to replace those containers. In truth, those containers and the tokens within them were superfluous. All that was necessary for the keeping of contracts was a systematic means of tracking goods and quantities in clay. Cuneiform writing, humankind's first writing, was quite likely born from this gradual realization. With time, the usage of this trade-based system for encoding quantities and goods was extended to other purposes. Novel symbols for goods and other concepts—ideograms—were developed time and again by new generations of scribes. The means of encoding these ideograms into clay was regularized, with rushes and reeds used to carefully inscribe symbols in an easily decipherable fashion. Grammatical features came to be conveyed as well, so that eventually any utterance could be depicted via writing. While we conveniently speak of the 'invention' of writing, it actually evolved over the course of millennia. It is clear, though, that at the start of this long evolution in Mesopotamia, representations of quantities already existed—they were in fact the core of early writing.

Something called the *rebus principle* accelerates the gradual evolution of writing systems, leading to the more rapid adoption of sound-based symbols from ideographic ones like the first forms of Sumerian writing. The rebus principle refers to the adoption of the same symbol in the representation of two homophones or similar-sounding words. For example, consider this fictitious but simple example: imagine if English writing were currently ideographic, so that a single symbol represented some basic idea or concept rather than a sound. And imagine further that the concept of an 'eye' were represented with the following configuration of parentheses with an asterisk in the middle: (*). This symbol could be said to represent in an iconic but somewhat abstract manner an actual eye. But say that there was not a symbol for the pronoun 'I' in this English writing system. One could imagine why, since there is no easy way to physically represent such a concept. Who 'I' is varies, depending on who is speaking, after all. If you are like many scribes in history, though, you might realize that you could use (*) to refer to both 'eye' and 'I,' given that they sound the same. This realization, referred to as the rebus principle, represents a major step toward a more abstract sound-based writing system with fewer symbols. Given their ancient status in writing systems, numerals themselves likely were affected by this principle, serving as symbols for homophones. This possibility is actually supported by modern-day forms of writing, for instance in texting. Somebody may acerbically text, for example, that they are "2 good 4 you." In such a case, the '2' and the '4' are used for texting ease, to more quickly and easily represent the phrase being communicated. One can easily imagine a scenario, however, in which abstract notions like 'too' and 'for' were represented by numerals, not because of ease of texting but because written forms of these very intangible notions did not yet exist.[11]

The rebus principle certainly acted as an accelerant in the development of syllable-based and letter-based systems for writing. But note that the rebus principle is dependent on the prior existence of symbols for less abstract things like goods and quantities. What is most remarkable in the context of our discussion is that the human tendency to encode quantities symbolically was more ancient and foundational, laying the groundwork for subsequent developments like the rebus principle, which in turn eventually led to writing systems like our own alphabet.

One of the forms of writing that eventually developed in Mesopotamia was written mathematics. Sumerians and the later residents of Mesopotamia, the Babylonians, developed elaborate written mathematical symbols. By 3,600 years ago, for example, Babylonians were already utilizing algebra and geometry, solving quadratic equations, and had already discovered π (at least approximately). So the representation of quantities seems to have ignited the development of Sumerian writing, which in turn eventually yielded the capacity to more elaborately represent quantities.[12]

In sum, the oldest writing system stems at least in part from the inherent usefulness of representing quantities, and from the relative ease with which numerical concepts can be abstractly represented. As we previously saw, this ease is also reflected in more ancient and less regular representational practices of the Paleolithic. From the stone age to the agricultural age, then, a lustrous thread of numbers winds through the human symbolic record.

Patterns in Ancient Numerals

While the Sumerians were apparently the first to employ full-fledged numerals, written numbers also evolved elsewhere. In fact, numerals have arisen at several points in world history. At least one hundred systems for writing down numbers have been documented, though the vast majority of these systems developed out of others or were at least developed with an awareness that other people already wrote down numerals. Many of the numeral systems in question are now defunct, though remaining examples permit us to understand how they worked.

When we examine extinct and existing numeral systems, we get a clear sense that there are common patterns in how humans write numbers. Let us consider these patterns by examining some numeral systems that have played a prominent role in human civilizations. The best place to start is with our own written numerals, commonly referred to as the Western numeral system. This system is a modified form of the Arabic notation system that is, in turn, a modified form of a system developed in India.[13]

So how does our system work? There are only ten symbols in our notation type: 0, 1, 2, 3, 4, 5, 6, 7, 8, and 9. This observation may seem glaringly obvious—of course there are only ten symbols. It may be difficult to conceive of numeral types with more or less than ten symbols, given the decimal nature of our own numerals. Yet notation systems need not be decimal and could have any number of symbols. One of the systems used by the ancient Greeks, their alphabetic numeral system, had about two dozen characters representing different values. Consider as well how we combine our ten number symbols to form larger numerals, say, two hundred and twenty-two: 222. Think of the conventions that are required for you to be able to make sense of that figure. You know that this sequence of 2s does not simply imply addition. 222 does not mean six, or $2 + 2 + 2$. The sequence implies multiplication. But the number is not merely the product of these three 2s either, or 222 would signify 8, or $2 \times 2 \times 2$. Instead, the 'places' or positions in our numeral system indicate an implicit multiplication by exponents of ten. So 222 signifies 2×10^2, plus 2×10^1, plus 2×10^0. That is, 200, plus 20, plus 2. In other words, when you read numerals transcribed in the Western numerical notation system, you are constantly adding together the products of numbers that have been multiplied by some exponent of ten. So 2,456,346 means to you $(2 \times 10^6) + (4 \times 10^5) + (5 \times 10^4) + (6 \times 10^3) + (3 \times 10^2) + (4 \times 10^1) + (6 \times 10^0)$. There is an inherent complexity to such numerals that is often overlooked because of their banality to us, and because of the apparent naturalness of grouping quantities into tens. But, despite the commonness of decimal systems in the world's numerals and in the world's spoken numbers, we are not somehow genetically hardwired to think of quantities in groups of ten. Learning our particular numeral system requires substantive effort, as evidenced by the amount of time it takes young children to acquire conventions for writing and reading large numerals. Furthermore, numeral conventions are culturally dependent, and a high degree of variability exists in the numerals that have developed in disparate regions. Many of these numeral types are not based on the grouping of quantities into tens. To illustrate this point, let me make a brief detour into the world of the ancient Maya.

Hidden behind dense tropical foliage and sometimes shrouded by mist, the stone city of Palenque was cloaked by nature for centuries prior to its "discovery" by Europeans in the latter half of the eighteenth century, long after the conquistadores and others had razed much of the cultural heritage of Central America. Ensconced in a sloping ridgeline at the periphery of the Chiapas highlands, this series of ruins has dazzled explorers since then. It was in Palenque that some of the earliest drawings of Mayan hieroglyphs were made by Europeans in the late eighteenth century, and it was there that some of the first photographs of those glyphs were taken in the late nineteenth century. While the Maya had produced innumerable codices (foldable bark-paper books with colorful writing), most of these were fodder for the pyres of Spanish friars, such as the infamous Fray Diego de Landa. With a view toward the forcible conversion of native Central Americans, de Landa extirpated much of the indigenous symbolic material culture and with it most examples of classical Mayan writing. He had native texts burned after learning that some Mayans still practiced their traditional belief system. As a result of his and others' subsequent efforts, only a few such texts or codices were preserved—three of which eventually made their way to European shelves in Paris, Madrid, and Dresden. The mysterious stone epigraphy in ruins like those of Palenque, Tikal, Copan, and other Mayan sites helped foster a fascination with Mayan writing. What did these exotic lithic representations of animals and ornately clothed humans, among numerous other symbol types, mean? Were they truly linguistic or primarily artistic in nature? Scholars hotly debated the answers to these and interrelated questions for decades, during the gradual decipherment of the Mayan script over the past two centuries.

Yet the first major breakthrough in the decipherment was made by someone who examined reproduced portions of the Mayan codex housed in the library of Dresden. In 1832, Constantine Samuel Rafinesque, an eccentric Frenchman with varied hobbies and talents, produced an analysis of patterns in the codex. The patterns he analyzed were also evident in stone engravings that had, already for decades, fascinated explorers of the palace at Palenque and other Mayan cities. What Rafinesque noticed was that, among all these engravings and writings, lost in a sea of seemingly undecipherable images, were recurrent series of dots and lines that were less iconic than the adjacent symbols. What could these lines and dots be? Rafinesque noticed that more than four dots were never placed in a row, but that arrays of one, two, three, and four dots were pervasive. In addition, he noticed that the dots were often placed adjacent to lines. He conjectured, and rightfully so, that the dots represented singular entities. One dot represented 1 element, two dots represented 2, and so forth. He inferred that a quantity of five was represented by a line, in a manner analogous to the way you might write a diagonal line through four marks when tallying five items on a piece of paper. This insight transformed our appreciation of Mayan symbols and was the first step toward deciphering Mayan writing. And a next major step also involved the Mayan numerals and revealed that the number system of these people was quite elaborate.

Decades after Rafinesque's epiphany, a German scholar named Ernst Förstemann offered analyses of the numerals depicted in the Dresden codex. Among other insights presented in several works published from 1880 to 1900, he observed that the Mayan numerals in the codex often depicted large quantities that corresponded to astronomical phenomena, such as the phases of Venus. Förstemann's rigorous work has held up to subsequent scrutiny and played a foundational role in the twentieth-century decipherment of Mayan glyphs. He detailed key components of the Mayan calendars and also shed light on the elaborate mathematics practiced by the Maya.[14]

While the Mayan system Förstemann helped decipher is not decimal based, it shares some clear structural correspondences with systems like our own. So just how did the Maya write numerals? For one thing, they used a symbol for zero—probably the oldest numeral for zero in the world. The Mayan symbol for zero is used as a place holder in their numerals, much as it is in our own. Unlike our horizontally oriented system in which digits are moved to the left one position to signify that they are multiplied by the next exponent of ten, Mayan numeral positions were stacked vertically to represent changes in exponents. (Though, confusingly, Mayan numerals could be rotated horizontally in texts.) Except, as Förstemann realized, the Mayan system of numerals was based on twenty rather than ten. In other words, it was a vigesimal system rather than a decimal one. In Figure 2.4 I have written two Mayan numerals to give you a sense of how the Mayan notation system worked. On the left is the representation for the number 437. The dotted lines in the figure help convey the vertical nature of Mayan numerals by separating the positions of the implied exponents, but such dotted lines were not actually present in Mayan writing. (Furthermore, for clarity's sake I have exaggerated somewhat the spacing between the positions.) In the number on the left, we see that the bottom-most series of lines and dots includes three lines and two dots, representing the number 17. We can think of this as $5 + 5 + 5 + 2$. In the middle portion of the glyph, there is only one dot. This dot represents 1 multiplied by the base raised to the first power. Since Mayan numerals were vigesimal, this means the dot represents 20

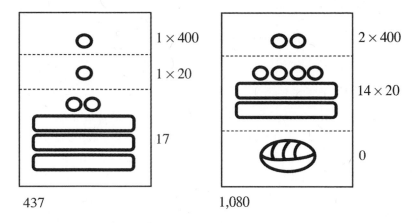

○	1×400	
○	1×20	
○○	17	

437

○○	2×400
○○○○	14×20
	0

1,080

2.4.   Sample Maya numerals. Note that, in calendric numerals, some dots could represent 360 instead of 400. This variant of the vigesimal system facilitated the tracking of years.

or 1×20^1. The topmost dot represents 1 multiplied by the base, raised to the power of 2, or 1×20^2. In other words, the topmost dot represents 400, the middle dot represents 20, and the bottom series of lines and dots represents 17. Together, these symbols represent $400 + 20 + 17$, or 437. Like our numerals, Mayan numerals involved multiplication and addition, but with a different base.

In the right portion of Figure 2.4 there is a depiction of 1,080. In this case the bottom-most portion of the numeral is filled with a variant of the oval-like Mayan symbol for zero. The middle series of lines and dots represents 14 (i.e., $5 + 5 + 4$). But the base-20 system of the Mayan numerals implied that this quantity of 14 was multiplied by 20, resulting in a total of 280. The topmost portion of the numeral contains two dots. These dots represent the quantity of 2×20^2, however, given their position. So, the two topmost dots represent 2×400 (800), the middle lines and dots represent 14×20 (280), and the bottom symbol represents 0×1 (0), yielding a total of 1,080.[15]

Mayan numerals may seem unwieldy, since they are not decimal based. But as I have mentioned, humans are not innately predisposed to think of items in groups of ten—most of us just happen to be well practiced with decimal groupings because of the language(s) we speak and the numeral system we are most familiar with. The Mayan system was quite functional and lasted for many generations and, furthermore, predates our Western numerals by many centuries. Had the Maya found the system unwieldy, they likely would not have utilized it for so long.

Despite the foreign-ness of the Mayan way of writing numerals, note that it shares much in common with our kind of numerals. First, the concept of positions is relied on heavily, as is a zero 'placeholder' number. While number positions are arranged vertically in the Mayan system and horizontally in our own, positions in both systems imply the multiplication of the displayed numeral by an implicit base and exponent. The base just happens to be ten in our system, and twenty in the Mayan. These bases are not random, though. The structuring role of five and twenty in Mayan numerals, and ten in our numerals, speaks to the manner in which the human body serves as the foundation of both numeral systems. It is no coincidence that humans have five digits on one hand, ten digits on both hands, and twenty digits on their hands and feet, and that these quantities play such an important role in Mayan and Western numerals. Furthermore, these quantities play a role in other independently innovated numeral types.[16]

Consider the Quipu system of the Incan empire. Quipu numerals were three dimensional, consisting of knots carefully tied into strings—strings made of cotton twine and other materials like the fur of alpacas and llamas. The strings were combined with other filaments, forming an array with a few or more than a thousand cords connected to a thicker main string. Each cord represented a separate number. Knots in the cords conveyed numerals and were used by Incan accountants, for instance, to keep track of taxes and goods and to conduct censuses. This system was unique in its form and material, but the symbolic composition of Incan numerals was actually quite similar to our own Western system. Quipu numerals were also based on the addition of multiplied groups of ten (i.e., they were decimal based), and also incorporated a means of denoting zero items. The number of knots on a particular cord corresponded to multiples of ten. Short gaps between knots were used to separate positions of a numeral, and longer gaps between knots represented zero. For instance, consider this series of knots on a string: one knot at the bottom of the string (1), followed by a slight gap of unknotted string, followed by three adjacent knots (30), followed by another short gap, followed by two adjacent knots (200), followed by a short gap, followed by one knot at the top of the string (1,000), near its intersection with the main, thicker cord of the quipu. Such a string would technically represent $1 + (3 \times 10^1) + (2 \times 10^2) + (1 \times 10^3)$, or 1,231. Or consider a string with one knot at the bottom (1), followed by a long gap of unknotted string (0), followed by three knots (300), followed by a slight gap, followed by two knots (2,000) at the top. Such a sequence of knots would correspond to $1 + (0 \times 10^1) + (3 \times 10^2) + (2 \times 10^3)$, or 2,301. In short, despite their obvious physical differences with respect to our Western numerals, quipu numerals were not entirely unlike our own, since they were decimal based and also relied on a kind of zero. This structural similarity is evident in the roughly 600 exemplars of the Quipu system that exist in museums and private collections.[17]

Most of the numeral types that have evolved over the past few millennia, across a variety of world regions, share some remarkable structural correspondences. Some of this overlap owes itself to the fact that written numbers only arose independently in a handful of locations. But not all the resemblances can be explained away as being due to the neighborly influence of surrounding cultures and empires. If so, a system like the Roman numerals might have had a greater impact on present-day notations. Instead that system, which lacked the concept of zero as a placeholder and frequently required long strings of symbols for relatively small quantities (e.g., XXXVIII for 38), eventually fell out of use except in restricted contexts. It gave way to a system that uses a symbol for zero and has positions that reflect groupings of ten.

Types of numerals vary in other ways not explored here. Nevertheless, the variation in the value of bases is remarkably constrained across unrelated numeral systems, both current and ancient. This limitation owes itself to a simple fact: The world's major numerical notation types, whether the decimal one invented in China and developed throughout eastern Asia, or those first used in India, Mesoamerica, or the Andes, share a common bias. They are all based somehow on ten or some other multiple of five. And the anatomical motivation for this bias is clear: regularly occurring quantities on our bodies are crucial to how we make numbers. This fact underlies written numerals as well as spoken numbers, as we will see in Chapter 3. Fingers and, to a lesser extent, toes have had a widespread influence in structuring numerals for millennia.

Conclusion

We have not comprehensively surveyed the history of human numeric artifacts in this chapter, and such an enterprise would require volumes. For example, we have not discussed the abacus systems that were and are used in various places, such as ancient Rome and contemporary Japan, to facilitate numeric thought, though I should mention that abaci around the world are also structured around groupings of five and ten.[18] But this survey has underscored some key points about ancient symbolic practices. First, it has demonstrated that the tabulation of quantities is evident in many Paleolithic artifacts around the world, suggesting that prehistoric numerals were some of the first nonverbal symbols (or less abstract quasi-symbols). Paleolithic art forms often depicted the fingers of a human hand, sometimes in configurations that are suggestive of ancient counting practices. Furthermore, the ease and usefulness of depicting quantities apparently made tallies and other prehistoric numerals pivotal to the first two-dimensional representations of ideas.

As systematic tabulation of larger quantities evolved, humans relied on the grouping of such quantities into smaller sets that were literally physically manageable. We eventually developed notation systems that were centered around particular naturally recurring sets of items—our fingers. A manual focus was crucial to the evolution of numerals, much as it was crucial to stone-age hand tracings on cavern walls. Our minds needed our bodies, in particular our fingers and hands, to keep track of quantities. This conclusion is well supported by spoken numbers, as detailed in Chapter 3. In this chapter we have seen that the conclusion is buttressed by distinct sorts of written numerals. We have also seen that numerals were present at the dawn of writing and likely were crucial to the development of the written word.

3

A NUMERICAL JOURNEY AROUND
THE WORLD TODAY

Anthropologists—whether archaeologists, linguists, or some other kind—traverse cultures, both in a physical and temporal sense, to learn about how people live and have lived. The overarching goal of our enterprise, arriving at a deeper understanding of what it means to be a member of the species *Homo sapiens,* depends on this cross-cultural inquiry. Ideally we are sponges, absorbing knowledge through direct interactions with other contemporary cultures or through indirect interactions, such as examining the remains of extinct cultures. In a sense, these interactions and exchanges are inevitably lopsided—what we take is usually of greater value than what we leave, no matter what sort of payment is involved.

Thankfully, we occasionally have the opportunity to share facets of our own cultures that are of value to those we are learning from. On a sweltering day a few Amazonian rainy seasons ago, such an opportunity presented itself to me. Playing soccer with a riverine group of Brazilians, I noticed we were joined by two indigenes, short in stature but with remarkable quickness and stamina. After the game (in which they were, not coincidentally, joint top-scorers) I struck up a conversation with them in their limited Portuguese. I discovered that these two were members of the Jarawara culture (mentioned in Chapter 1), an indigenous group of about one hundred people. Immediately I was quite keen to learn from these men, since they are speakers of one of the few languages that has been claimed to have no number words. I soon discovered they were also keen to acquire something from me—knowledge of how to ride the motorcycle on which they had seen me arrive. Over the next several weeks, we exchanged these components of our native cultures with each other: I learned about their language, and they learned how to ride an off-road motorbike. During the time spent with them and some of their friends and family (who like them were on an excursion away from village life), I realized two main things: first, contrary to previous claims, the people do have a fascinating native number system, and second, they are adept and fearless when it comes to learning how to ride a motorcycle.

The Jarawara live in two principal villages located near the Purus River, a main fluvial arm of the Amazon. They speak one of a cluster of related languages in this region, descendants of an extinct language termed 'proto-Arawa,' which was likely spoken more than 1,000 years ago. We now know that all these related languages have a few basic number words that bear resemblance to one another. For instance, the words for 'two' in all the Arawa languages are similar, suggesting the words are what linguists refer to as *cognates*. Cognates derive from the same word in an ancestral language, rather than being borrowed across languages more recently, and they help linguists reconstruct the extinct words they descend from. In the case of proto-Arawa, we are now confident that the word for 'two' was *pama*. (In linguistics an asterisk denotes a reconstructed word that existed in an ancestral language.) In most language families of the world, we can reconstruct ancient words for numbers, such as *pama, since the vast majority of the world's current languages have words for precise quantities. Even Australian languages, which generally lack large inventories of number words, have words for some quantities. All of this suggests that spoken numbers are an extremely ancient human innovation, common to the world's current languages as well as to long-ago spoken ancestral tongues. In this chapter we survey some of the key findings evident in the world's spoken numbers, which I refer to simply as numbers. (In contrast to 'numerals,' a term I reserve for written numbers.)[1]

The word for 'two' in Jarawara, it turns out, is *fama,* clearly similar to the *pama word used by proto-Arawa speakers. During interviews with the Jarawara acquaintances I made, it became clear that their native number system actually has many numbers beyond *fama* as well. In individual meetings with them, I asked them to provide a word from their language to describe the quantity of an array of items I placed before them on a table. The seven adult speakers that volunteered to help me were consistent in their responses. Furthermore, when subsequently asked to provide translations for Portuguese number words, the speakers were generally consistent again. As a result of these sessions, I concluded that Jarawara has a native number system, contrary to previous claims. As in many other languages in the world today, native number words are being replaced in Jarawara with the more useful numbers of the hegemonic economic power the people are forced to interact with. (See the discussion of endangered number systems in Chapter 9.) Despite their adoption of the Portuguese numbers used throughout Brazil, however, the Jarawara adults I met could still recall their traditional cardinal number words.[2] Some numbers in Jarawara are shown here (parenthetical words are optional):

QUANTITY	JARAWARA TERM
1	ohari
2	fama
3	fama oharimake
4	famafama
5	(yehe) kahari
7	(yehe) kahari famamake
10	(yehe) kafama
11	(yehe) kafama ohari
20	(yehe) kafama kafama

The Jarawara number system is a useful gateway to our survey of numbers in the world's spoken languages, because the patterns evident in the system are indicative of commonalities shared by most of the world's languages. In particular, the bases of Jarawara numbers are evident in many of the world's number systems. A base is a word that recurs—usually explicitly though perhaps only implicitly—in the numbers of a given language. A base is a building block for other numbers.[3] (As we saw in Chapter 2, the term can also refer to the value raised to specific powers in a written numeral system.) Let us consider briefly the bases of Jarawara numbers. First, one notes the recurrence of the word *fama* throughout the numbers. So, while 'two' is *fama,* 'four' is *famafama,* a reduplicated form of the word 'two.' Similarly, the word for 'ten' is *(yehe) kafama,* meaning "with two (hands)", wherein the word for hand, *yehe,* may be implicit rather than explicit. We can conclude, then, that Jarawara has what is known as a binary base (base 2). That is, the number 'two' is used as a building block for at least some larger numbers. Yet this is clearly not the only base evident. Beginning with the number 'five,' we observe that *yehe* is used, at least implicitly, in all the remaining numbers. The word for 'five' itself is best translated as "with a hand." This quinary base (base 5) of the number system is evident throughout. The word for 'seven' is best translated as "a hand with a pair," and the word for 'ten,' *(yehe) kafama,* is translated as "with two hands." This word for ten is then used as a base for the remaining numbers, such as 'eleven' and 'twenty.'

Clearly, the Jarawara number system has recurrent bases rather than original names for each named quantity. In this sense it typifies how most verbal number systems work. It also typifies many number systems in that the number five has a special status as a recurrent base. Additionally, the number ten recurs, though it is itself based on the number five. Most typically, the world's verbal numbers are decimal—based on the number ten. Yet they are frequently quinary as well, as is the case in Jarawara. Furthermore, they are sometimes vigesimal, based on the number twenty. Finally, the binary base evident in Jarawara also occurs in some other languages of the world. In short, this recently uncovered number system is fairly representative of the number systems evident in the world's languages. Certain words are used as building blocks in spoken numbers, and these 'bases' tend to denote such quantities as 5, 10, 20, and, less frequently, 2. Framed differently, numbers reflect a strong bias toward our construal of quantities through individuated units of our biology, principally our fingers.

Consider another Amazonian example, the language of the Karitiâna people that has been the subject of some of my own research. This language is completely unrelated to Jarawara. As in most of the world's number systems, the terms for smaller quantities in this language are un-analyzable, without any identifiable bases like the binary *fama* of Jarawara. In the list below I present just a handful (it is difficult to escape the hands when talking about quantities, even in English) of Karitiâna number words.

QUANTITY	KARITIÂNA TERM
1	myhint
2	sypom
3	myjyp
4	otadnamyn
5	yj-pyt ("our hand")
6	myhint yj-py ota oot ("take one and our other hand")
11	myhint yj-piopy oot ("take our one toe")

The patterns in this list are illustrative, again, of commonalities in the world's number words. The word for 'five' is clearly associated with the word for 'hand' and, furthermore, it serves as a base for higher numbers, such as 'six.' In addition to their quinary foundation, Karitiâna numbers reflect the decimal base that is more common in the world's languages. Consider the word for 'eleven,' best transliterated as "take our one toe." Such a number word is implicitly based on 10, since only one human digit, a toe, need be referred to in order to denote 11. The same could be said for other higher numbers in Karitiâna, which are also decimal based.

In fact, a great majority of the world's number systems, across all of its roughly 7,000 languages, reveal decimality in one form or another, since higher numbers are most commonly structured around 10. Lower quantities are frequently quinary based (as in the case of Jarawara and Karitiâna) or unanalyzable (as in the case of the smallest Karitiâna numbers). The motivation for this quinary and decimal fixation by human number creators has been acknowledged for some time—the world's peoples tend to rely on their hands when coming up with number words. Our fingers allow for the simplest extension of numerical concepts into the physical world. That extension, reified through finger counting and associated practices, is then further extended into the verbal realm, via the naming of quantities through the metonymic transfer of biological terms like 'hand,' 'finger,' and 'toe.'

In many languages, higher numbers are phraselike, with obvious physical origins, as in the Karitiâna word for 11, "take our one toe." One of the tenets of modern linguistic theory is that terms that are used frequently are reduced phonetically—commonly used words get shorter with time. Compound words and other once-new phrasal words are shortened. For this and other reasons, frequently used words tend to have less obvious historical sources. Less frequently used words are more likely to retain their original form. In a sense, the obvious sources of some numbers in languages like Karitiâna and Jarawara reflect their relatively infrequent usage. In contrast, in most languages in which numbers play a more prominent role and are more frequent in speech, even higher numbers have less obvious etymologies. (Recall that Jarawara numbers are used so infrequently that some have assumed they do not exist.) In English, for instance, higher number words do not involve any obvious reference to fingers, hands, toes, or feet.[4]

Any *obvious* reference. But less obvious references are there. Because English number words, apart from the way they are written down in numerals, are rigidly decimal. Consider the words 'thirteen,' 'fourteen,' and any other word ending in 'teen.' These words indicate that quantities are being added to ten. 'Thir+teen' is, simply, 13, 'four+teen' is 14, and so on. Numbers beyond teens are also decimal, though the concept of 10 has taken the shape of 'ty' in such words, rather than 'teen.' Furthermore, these higher numbers are based on multiplication rather than addition. But 'twenty,' 'thirty,' 'forty,' 'fifty,' 'sixty,' 'seventy,' 'eighty,' and 'ninety' clearly reflect simple math problems centered on the quantity 10: 2×10, 3×10, 4×10, and so on. This decimal fixation has the same biological roots as it does in a language like Karitiâna. People count with their fingers, and sometimes with their toes, and when they name quantities, the names chosen are based somehow on the identification of those quantities with human digits.

This pattern is pervasive in the world's languages. Consider another example of a decimally oriented European language, Portuguese. While it is part of the large Indo-European family of languages, Portuguese is a Romance language. So it is not particularly closely related to English, a member of the Germanic branch of the Indo-European tree. In some cases, number words in the two languages bear resemblance to each other, in others not. Setting aside any terminological disparities, though, the decimally oriented nature of Portuguese is also clearly evident:

QUANTITY	PORTUGUESE TERM
1	um
2	dois
3	tres
4	quatro
5	cinco
6	seis
7	sete
8	olto
9	nove
10	dez
11	onze
12	doze
13	treze
14	quatorze
15	quinze
16	dezesseis
17	dezessete

18	dezoito
19	dezenove
20, 21, 22, . . .	vinte, vinte um, vinte dois, . . .

Again, we see that the forms of the words for 1–10 are seemingly arbitrary. Yet beginning with 11 we see some regularity. 'Eleven' is the number at which counting starts over, so the preceding number must serve as a base. For numbers 11–15, we can observe a recurring vague resemblance to numbers 1–5 plus the form *ze*, which is now meaningless by itself but clearly reflects the ancient addition of 10 to 1–5. Similarly, for numbers 16–19, we see they resemble numbers 6–9, but once again these are combined with 10. In this case the combination is more transparent, since *e* is the conjunction word 'and' in contemporary Portuguese. So 16, for instance, is literally "ten and six." For higher numbers, the system once again regenerates at the next multiple of ten, so that 21, for instance, is *vinte um* or 'twenty one.' The same is true of 31, 41, and so on. In short, Portuguese numbers are clearly decimal.[5]

It is worth stressing that the decimal nature of these number systems is not due simply to some mathematical expediency, as is sometimes assumed. Decimal bases are evident in many societies without mathematical traditions, like Karitiâna, and, in European languages, they predate by millennia the advent of written numerals and modern mathematical practices. This is evident in the reconstructed number system of Proto-Indo-European, the ancestor of English, Portuguese, and more than 400 other related languages.[6] Proto-Indo-European was spoken about 6,000 years ago, somewhere in the neighborhood of the Black Sea. (Specialists still hotly debate exactly where it was spoken, either in the steppes of present-day Ukraine or in Anatolia.) Through some series of events, speakers of Proto-Indo-European and its descendant languages came to influence the world in an unparalleled fashion, and nearly half of the world's people today speak an Indo-European language of some kind. Some reconstructed numbers from this influential ancestral language are presented below. Recall that an asterisk denotes reconstructed forms. That is, we have no transcripts or historical confirmation of these words, but based on similarities among modern languages of Indo-European descent, we can be fairly confident that the number terms used in the language took the form presented.[7]

QUANTITY	PROTO-INDO-EUROPEAN TERM
1	*Hoi(H)nos
2	*duoh
3	*treies
4	*kwetuor
5	*penkwe
6	*(s)uéks
7	*séptm
8	*hekteh
9	*(h)néun
10	*dékmt
20	*duidkmti
30	*trihdkomth
40	*kweturdkomth
50	*penkwedkomth
60	*ueksdkomth
70	*septmdkomth
80	*hekthdkomth
90	*hneundkomth
100	*dkmtom

The decimal nature of Proto-Indo-European numbers is obvious, regardless of minor discrepancies in alternate reconstructions. Numbers 11–19, not presented in the list above, take some variation of 'ten and x,' where x is some number from 1 to 9. Numbers for 20 and beyond also clearly reflect a base-10 pattern. These numbers contain some variant of the *dkmt* string of sounds, owing to the ancient number for 10, **dékmt*. Rather than being additive, however, they are multiplicative—just as we saw in English and Portuguese. So 20, for instance, is **duidkmti,* or (roughly) 'two tens,' while 30 is **trihdkomth,* or 'three tens.' And so on.

The ancient status of decimal bases in spoken numbers is also evident in other major families of the world's languages. Let us consider three more of these: Sino-Tibetan, Niger-Congo, and Austronesian. As with Indo-European, there are more than 400 Sino-Tibetan languages spoken today, and their speakers number well over 1 billion, as these languages include Mandarin and Cantonese. Since a discussion of numbers in Proto-Sino-Tibetan would necessitate an esoteric outline of phonetic aspects of that language, let us look at numbers in the most widely spoken language of this stock of languages today: Mandarin. As with European languages, Mandarin has undecipherable numbers for 1–10. That is, there are no recurring units used in the construction of these smaller numbers. The Mandarin word for 10 is *shí*. Words for 11–19 all contain this word, plus some number that is also used to represent a number less than ten. The word for 11, for instance, is *shíyī,* or 'ten one.' The word for 17 is *shíqī,* or 'ten seven.' Higher numbers are also decimal based, reflecting a grouping of quantities according to ten items. In the case of numbers for 20–99, however, smaller numbers come before the decimal base. So 70 is represented with *qīshí,* or 'seven ten,' in direct contrast to the reverse pairing evident for 17. The same holds for other multiples of ten. In short,

Mandarin numbers have a decimal base, in fact one that is more transparent than in the case of English.

Niger-Congo is another of the world's major language families. In terms of number of languages, this is the world's largest linguistic grouping, with more than 1,500 representatives according to one recent survey. One of the most widely spoken Niger-Congo languages is Swahili, a language of the Bantu branch used extensively in eastern Africa. The decimal basis of Swahili, indicative of the decimal strategies prevalent throughout the Niger-Congo family, is apparent in the examples provided below, which demonstrate that quantities like 11–13 are conveyed by adding terms for smaller numbers, such as *tatu* or 'three,' to the word for 'ten,' *kumi*.

QUANTITY	SWAHILI TERM
1	moja
2	mbili
3	tatu
11	kumi na moja
12	kumi na mbili
13	kumi na tatu

The Austronesian language family is incredibly dispersed geographically, with more than 1,200 representatives in places as far away from each other as Madagascar, southeast Asia, and Hawaii. Austronesian languages are scattered throughout many other Pacific islands as well, due to the seafaring expansion of speakers of ancestral Austronesian tongues over the past 2,000 years or so. Proto-Austronesian apparently had a decimal system, and most Austronesian languages are characterized by a decimal base. This is true, for instance, in the branch of the language family known as

Polynesian, which has been shown to have a longstanding decimal system.[8] Interestingly, many Austronesian languages also show evidence of a quinary base. Like the decimal base, this quinary pattern has undeniable anatomical motivations, particularly since the word for 5 in proto-Austronesian is the same as that for 'hand': *lima.[9]

Our numerical journey has so far served to demonstrate that the world's largest language families, including Indo-European, Sino-Tibetan, Niger-Congo, and Austronesian, like smaller language families in Amazonia and elsewhere, reveal the pervasive influence of decimality. This influence clearly stretches back many millennia. Human cultures have long since arrived independently at the notion that numbers are naturally grouped by tens and fives. As we saw in our discussion of written numerals, this cognitive habit is due to our biology. Number words are often sourced from words for hands. Even when they are not, the decimal (and, to a lesser extent, vigesimal and quinary) bases in most number systems reflect the anatomical basis of lexical numbers. In a recent survey of 196 languages from diverse families and geographic regions, renowned linguist Bernard Comrie found that 125 of them had pure decimal bases for quantities greater than ten. Hybrid decimal / vigesimal bases and pure vigesimal bases were also found to be common, with 22 cases of the former and 20 of the latter present in the sample.[10]

The vigesimal base is relatively prevalent in such regions as central America, the Caucasus, and west-central Africa. Yet vigesimal patterns are also weakly evident in some European languages. They are particularly apparent in French, where the structure of many higher numbers reflects a base-20 pattern. For instance, the word for 99, *quatre-vingt-dix-neuf* ($4 \times 20 + 10 + 9$), is founded on vigesimal and decimal bases. Even English has some infrequently evident vestiges of a base-20 system. Consider Abe Lincoln's famous introduction to his address at Gettysburg, "Four score and seven years ago," in which he referenced the independence of the United States some 87 years prior to his speech. The fact that the word 'score' exists, no matter how arcane it is, reflects a vigesimal linguistic grouping of quantities.

In some cases, a single language may exhibit quinary, decimal, and vigesimal bases. In Mamvu, a language of the Central Sudanic family in Africa, the words for 1–5 have opaque etymologies. In other words, as is so commonly the case, the sources of the smaller number words are mysterious. In contrast, words for 6–9 are quinary based. The word for 6, *elí qodè relí,* is best translated as 'the hand seizes one.' The phrasal number *qarú qodè relí,* however, means 'the foot seizes one' and signifies 11. In contrast, words for quantities beyond 20 are based on the phrasal number *múdo ngburú relí,* meaning 'one whole person.' Languages like Mamvu represent particularly clear evidence that number systems are generally constructed around units referring to one hand, two hands, and whole persons, where the persons are understood to refer to all the fingers and toes of an individual.[11]

Despite the commonness of these bases in the world's languages, it should be underscored that the decimal base in particular plays a prominent role in numbers. The prominence of the decimal base is clearly due to human biology rather than, as is sometimes assumed, some inherent roundedness or efficacy of grouping things into 10. And the motivation for base-10 prevalence when compared to base-20 is also clearly biological: our fingers are much more salient and accessible than our toes. They are in our field of vision, more easily manipulated, and serve as a more natural means of counting items. This usage in counting routines helps motivate the naming of number terms. The motivation for humans' greater reliance on base-10 systems rather than base-5 ones is perhaps a little less obvious but is once again explainable via simple anatomical facts: fingers are relatively homogeneous, across both hands, however there is a clear distinction between the fingers and toes. This is true with respect to their locations on the body and their physical appearance. As linguist Bernd Heine puts it, "since the perceptual difference is larger between hands and feet than between one hand and another, the numeral '10' appears to constitute a more salient base than '5'."[12]

As is apparent in some of the examples we have considered, number words for quantities beyond five are often phrasal in nature. In Karitiâna, for instance, 6 is represented with the phrasal number 'take one and our other hand.' Such phrases often lexicalize with time (i.e., they become shorter words with less obvious meanings). This is what has happened in the case of English numbers. Yet, in some languages these phrasal numbers remain transparent and consist of additive or multiplicative expressions. As we saw in the case of Mandarin, numbers for 11–19 are additive, meaning 10 plus x, while numbers for 20, 30, 40, and so forth are multiplicative, taking the form of x times 10. Most number systems implicitly rely on addition and multiplication to construct larger numbers from smaller ones, as in Mandarin. Much less frequently, subtraction is used, even in languages that are also based on decimal groupings. In the case of the Ainu language indigenous to Japan, the word for 9 is *shine-pesan,* meaning 'one down,' and the words for 6–8 are also subtraction based. Subtraction is never the primary basis for creating numbers in a given language, however, and there is a hierarchy evident in the world's number systems wherein addition plays a more prominent role than multiplication, which in turn plays a more prominent role than subtraction. Division is used as well, but in exceedingly rare instances.[13]

Along with the primacy of addition in the construction of number words, another associated tendency is evident in many phrasal numbers in the world's languages: addition is denoted via the common usage of words such as 'take' or 'seize.' As we saw in

Karitiâna, the word for 11 is translated 'take our one toe.' In Mamvu the word for 11 is translated as 'the foot seizes one.' Along with words for 'take' and 'seize,' another common way smaller numbers are added to form larger number terms is through words or morphemes meaning 'with.' This is the case in Jarawara, in which the prefix *ka-* is used in an expression like *yehe kafama,* literally "with two hands" or 'ten.'

The body-part basis of number words is also apparent in a few other weaker trends in the world's languages. For instance, words for the act of counting are often derived historically from words for fingers or for the bending of fingers. (The latter association between counting words and words for the bending of fingers is evident, for example, in languages of the northern branch of the North American Athabaskan family, and in Zulu.)

Other generalizations, unrelated to the body-part basis of numbers, could be made about the typical forms that numbers take. We have not discussed, for instance, very large numbers like 'thousands' and 'millions,' which also reflect an ultimately decimal basis, as they are based on 10 being raise to particular powers, even if they have different etymological sources when contrasted to smaller numbers. These larger numbers also differ from smaller ones in that they tend to be treated as nouns, unlike smaller numbers that tend to be treated as adjectives. For example, in English I can talk about 'hundreds' or 'thousands' of people but I cannot say there were 'sevens' or 'eights' of people. Furthermore, these larger numbers are comparatively rare in actual speech, and are less common in the world's languages as well. While nearly all languages of the world have a system of basic numbers,[14] these often do not have the means of conveying quantities beyond 100, or 20, or some other multiple of 10 or 20. In fact, as we shall see in the following section, the ceiling on the precise native number terms of many languages in Australia, Amazonia, and elsewhere is less than ten.

From this brief survey we can reach a few broad conclusions about numbers in speech. First, they are common to nearly all the world's languages. While we have only examined a small fraction of potential examples from those languages, the examples illustrate that basic number words are common to unrelated languages in far-flung geographic regions. Second, this discussion has shown that number words, across unrelated languages, tend to exhibit striking parallels, since most languages employ a biologically based body-part model evident in their number bases. Third, the linguistic evidence suggests not only that this body-part model has motivated the innovation of numbers throughout the world, but also that this body-part basis of number words stretches back historically as far as the linguistic data can take us. It is evident in reconstructions of ancestral languages, including Proto-Sino-Tibetan, Proto-Niger-Congo, Proto-Austronesian, and Proto-Indo-European, the languages whose descendant tongues are best represented in the world today.

Our bodies play a prominent role in our thought patterns, in how we make sense of the world around us. This physical influence on thought is evident in much current research in cognitive science on the way that the body shapes our cognitive processes, research on so-called embodied cognition. Basic quantitative reasoning is an example par excellence of embodied cognition, since we make sense of quantities through our bodies. (A point taken up further in Chapter 8.) As we saw in Chapter 1, humans seem to have a hard time making sense of the passing of time unless they enlist their bodies, and the space around those bodies, as a concrete metaphorical basis for thinking about time. In a similar manner, our survey of basic number words suggests that the primary way we make sense of abstract quantities is through our fingers and toes, but primarily through our fingers.

In some respects this all may seem obvious. After all, we all used our fingers to count in our youth, and sometimes we still do. Yet the physical basis of numbers is frequently underacknowledged, I think, particularly in the light of the ubiquity of the body-part model in the world's spoken numbers. This ubiquity suggests that number words are inventions—invented out of an association between our fingers and the quantities they can be used to represent. (This point is discussed more fully in Chapter 8.) Alternatively, one could maintain that numbers are independent concepts, perhaps innate in the human mind, and that we simply use our fingers to label these concepts. As we shall see in Chapter 5, such a position is difficult to maintain in the face of experimental evidence with anumeric populations. Yet it is worth noting here that the finger-centric nature of number words in the world's languages should also give pause to those believing that numbers are concepts that merely await labeling. Surely if that were the case, we would not rely on our fingers so often to label them, and the world's languages would contain all manner of alternate bases that would not imply the historical usage of fingers. The crosslinguistic evidence alone more neatly supports the perspective that numbers are conceptual tools, not just labels, created after we make sense of quantities with our fingers.[15]

The invention of numbers is attainable by the human mind but is attained through our fingers. Linguistic data, both historical and current, suggest that numbers in disparate cultures have arisen independently, on an indeterminate range of occasions, through the realization that hands can be used to name quantities like 5 and 10. Fingers can be used to correspond to quantities, and many people have discovered the utility of doing so. Yet fingers are limited symbolically and do not represent a fully abstract means of conveying quantities. Words, our ultimate implements for abstract symbolization, can thankfully be enlisted to denote quantities. But they are usually enlisted only after people establish a more concrete embodied correspondence between their fingers and quantities. Our invention of numbers is grounded in our bodies, then, in the digits that have enabled us to derive truly abstract symbols for quantities, number words. These abstract symbols can be easily learned and transferred in and across populations, and have come to serve a host of needs. They are true verbal and conceptual tools.

Other Motivations for Quantity Groupings

Despite the global prevalence of number systems based implicitly or explicitly on the human digits, numbers may also be based on other factors. In addition to quinary and decimal systems, there are also binary ones. In the case of Jarawara, for instance, we saw some binary influence for smaller numbers. In the traditional number system once utilized on the island of Mangareva, a more elaborate sort of binary system was employed, as we shall see in Chapter 9. In addition to the binary or base-2 numbers that exist in some cultures, other bases evident in more than one culture include ternary (base 3), quaternary (base 4), senary (base 6), octonary (base 8), and nonary (base 9). Additionally, some languages utilize or have utilized a duodecimal (base-12) system and others a sexagesimal (base-60) base. As was noted in Chapter 1, the ancient usage of a base-60 system, first by Sumerians and later by Babylonians, is the reason we still divide hours into minutes and minutes into seconds.

Interestingly, the octonary, duodecimal, and sexagesimal bases may stem from the human hands as well, though in less obvious ways than quinary or decimal groupings. While humans have ten fingers, not all observed finger-counting methods are based simply on a correspondence between fingers and quantities. There are ways to represent quantities via other regularities of the hands. Due in part to such alternate though less obvious regularities, there is certainly cross-cultural variation with respect to how people count with their fingers. To cite one exotic example, in an Indian merchant finger-counting system the fingers on one hand represent individual units, while those on the other hand represent multiples of five. So two fingers on one hand plus three fingers on the other hand would represent 17 ($2 \times 1 + 3 \times 5$), not 5.

The spaces between the fingers are also easily countable. In fact, the rare base-8 (octonary) system may be motivated by the total number of spaces between fingers on our two hands. Or consider the base-12 or duodecimal pattern. There are twelve adjacent lines on the non-thumb digits of each hand (one for each knuckle), and these lines can serve as salient marks for denoting quantities, counted off by a finger from the other hand. (See Figure 3.1.) Furthermore, beyond this potential biological basis for the base-12 strategy, it is worth noting that $12 \times 5 = 60$. Given that base-12 and base-5 strategies each have a clear potential motivation in the human hands, it is less startling that a base-60 system arose in ancient Mesopotamia. This is not to suggest definitively that a base-60 strategy is manually based. Yet it certainly could be, and the usage of a base that is neatly divisible by 5, 10, and 12 seems an unlikely coincidence.[16]

I do not wish to belabor the particulars of all these less common sorts of number bases, but it is vital to note their existence so that I do not give the impression that humans are somehow required, via some innate mechanism or some other factor, to make sense of quantities through our fingers in the exact same fashion. Furthermore, we can and do use other items in our environments, most frequently other features of our own biology, to make sense of quantities. Binary systems may be due in part to the fact that human biological features are in many cases structured in pairs. In particular, our heads have salient pairs of eyes, ears, nostrils, and cheeks. In some languages there is some weak evidence that the word for 'two' derives from one of these words. For instance, the word for 'two' in Karitiâna is *sypo*, and the word for 'eye' is *sypom*. (It is unclear whether this fact is coincidental.)

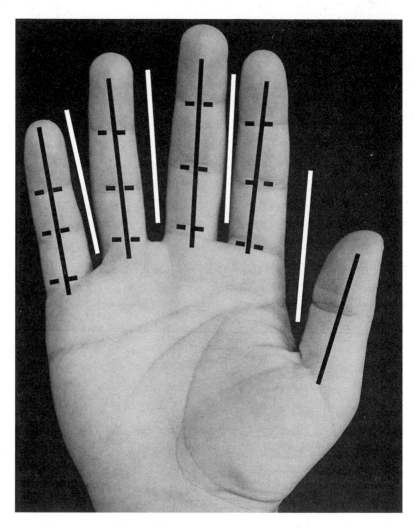

3.1.   The majority of the world's number systems are based on the quantity of fingers of the human hands. However, other features of our hands may also influence number systems. The four gaps highlighted with white lines and the twelve adjacent joints highlighted with dotted black lines play less common roles in shaping numbers. Photograph by the author.

Other body-based counting strategies, sometimes idiosyncratic ones, also exist. In the West Sepik province of New Guinea there is a language, Oksapmin, which once utilized a counting strategy wherein 27 numbers corresponded sequentially to parts of the body, including the fingers, eyes, and shoulders. The countable points on the body began with one hand, proceeded up one arm, over the head, and then down the other arm to the other hand. Similar systems have been documented elsewhere in New Guinea, for instance that used among the Yupno. In their language, counting terms are associated with the fingers, toes, as well as other points on the body, yielding a sequence of 33 words. Since they do not lead to regular number bases in speech, we do not discuss such number systems further here. Yet their existence is worth noting as they suggest yet another way in which the structure of the human body motivates counting strategies.[17]

Rare number bases have been observed, for instance, in the quaternary (base-4) systems of Salinan languages of California, or in the senary (base-6) systems that are found in southern New Guinea. Senary systems have received a fair amount of attention from language researchers. One of the results of that attention is the claim that numbers are senary-based in some languages of New Guinea because of a facet of regional culture. Yams play a fundamental role in the economies and subsistence of the groups indigenous to this part of the world. Fascinatingly, the six-by-six arrangements that are typically used to group and store yams seem to have motivated the senary number systems employed in the region. This object-specific counting strategy was at some point made more abstract and extended to other domains, so that all items are countable via the base-six system.[18]

Other number types are also influenced by counting strategies associated with objects external to the body, objects that play an important role in the cultures of the languages' speakers. Several languages in Melanesia and Polynesia have or once had number systems that vary in accordance with the type of object being counted. In the case of Old High Fijian, for instance, the word for 100 was *bola* when people were counting canoes, but *koro* when they were counting coconuts. While the number base was still decimally influenced, some object specificity was apparent. Polynesian numbers will be discussed more in Chapter 9, where we will see that that they present some cognitive advantages.

Intriguingly, some rare number systems reflect the influence of more esoteric cultural phenomena. As University of Texas linguist Patience Epps has discovered, some languages in northwest Amazonia base their numbers on kinship relationships. This is true of Dâw and Hup, two related languages in the region. Speakers of the former language use fingers complemented with words while counting from 4 to 10. The fingers signify the quantity of items being counted, but words are used to denote whether the quantity is odd or even. If the quantity is even, speakers say it "has a brother," if it is odd they state it "has no brother." Similarly, in Hup the word for 3 is best translated as "without a sibling" and the word for 4 as "have sibling, accompany." This 'fraternal' number system apparently stems from the sibling-exchange method of marriage practiced in this part of South America. Like the yam-based senary numbers in New Guinea, these relationship-based numbers demonstrate that just because numbers are body based in most cases does not mean they have to be. As with other aspects of language, we can only claim that common but nonuniversal patterns exist in the world's number systems.[19]

Limited Systems of Spoken Numbers

Since Dâw numbers 4–10 require the use of fingers along with 'brother' references, they are not verbal numbers in the strictest sense. Dâw only has true lexical numbers for 1–3. Like Dâw, some other languages in the world have numbers that are limited in scope. Some of these languages have base-less number systems. In one recent survey of limited number systems, it was found that more than a dozen languages lack bases altogether, and several do not have words for exact quantities beyond 2 and, in some cases, beyond 1. Of course, such cases represent a miniscule fraction of the world's languages, the bulk of which have number bases reflecting the body-part model. Furthermore, most of the extreme cases in question are restricted geographically to Amazonia. Also, it is difficult to state with absolute confidence that bases never existed in some of these supposedly base-less languages, since in some instances native numbers may have fallen into disuse due to the adoption of a more productive number system. (Recall our discussion of Jarawara.) In some cultures, younger speakers may fail to acquire the older native number system, and the extinction of that number system could give the incorrect impression that one never existed. Even in the light of this sociolinguistic factor, however, we can be confident that some languages have limited native numbers without bases. In the case of two Amazonian languages, Xilixana and Pirahã, it has been claimed that there are no precise number words of any kind. The case of Pirahã, which I am familiar with, is taken up in some detail in Chapter 5. In the less clear-cut case of Xilixana, some reports have suggested that the three numbers in the language refer, vaguely, to 'one or a few,' 'two or a few,' and 'three or more.'[20]

Some of the restricted number systems in Amazonia allow for exact reference to 1 or 2 items, but only for imprecise reference to greater quantities. This characterization is true of the Munduruku language, which has received much attention in recent psycholinguistic research. (See Chapter 5.) Famously, most of the languages of Australia have somewhat limited number systems, and some linguists previously claimed that most Australian languages lack precise terms for quantities beyond 2. That claim seems to have been overstated, however, and many languages on that continent actually have native means of describing various quantities in precise ways, and their number words for small quantities can sometimes be combined to represent larger quantities via the additive and even multiplicative usage of bases. As noted in an extensive survey of Australian languages conducted by linguists Claire Bowern and Jason Zentz, the number systems are in fact frequently limited. Yet they certainly cannot be described as "one, two, many" systems, as was once conjectured. They tend to be more productive, for instance, than the number systems of hunter-gatherer groups in Amazonia. Of the nearly 200 Australian languages considered in the survey, all have words to denote 1 and 2. In about three-quarters of the languages, however, the highest number is 3 or 4. Still, many of the languages use a word for 'two' as a base for other numbers. Several of the languages use a word for 'five' as a base, and eight of the languages top out at a word for 'ten.' (Whereas there are no languages that refer to, say, 7, 8, 9, or 11 with their largest number.) Even in Australia, then, there are scattered hints of the finger foundations of number-making strategies. This is particularly telling, given the relative isolation of the peoples on that continent since its settlement more than 40,000 years ago. As in other world regions, then, some Australian populations have independently innovated the body-part model for representing quantities linguistically.[21]

Conclusion

Where once most language researchers considered linguistic differences to be relatively superficial, obscuring the deeper, universal characteristics of all languages, a growing number of them now consider the diversity of extant human languages so profound that it is empirically unjustifiable to speak of any meaningful features universal to all languages.[22] In our short foray into the world's spoken numbers, we have seen that there certainly are no universals to find here, contravening old expectations that all languages have precise number words. Furthermore, we have seen that languages vary in terms of how they construct number systems, with some of them being quite limited in scope and others, like our own, potentially infinite. We have also seen that indigenous languages sometimes have more numbers than initial impressions suggest, for instance in Australia or among the Jarawara. Careful linguistic documentation is yielding a more accurate picture of the world's numbers.

Yet, superimposed over this crystallizing picture of numerical diversity are clear patterns evident in the world's spoken numbers. The patterns are simple: number words are frequently derived from words for hands and typically reflect the grouping of quantities into tens, or fives, or twenties, or some combination thereof. These simple facts, evident in ancient and contemporary languages on every inhabited continent, suggest that number bases are verbal reifications of concepts made clearer to us through the embodiment of quantities. This embodiment is made possible because of naturally occurring sets of items so salient to us, literally at our fingertips just waiting to be grasped. (In Chapter 8 I offer a more specific account of how this 'grasping' happens.) Our discrete and match-able fingers have enabled us to concretize abstract and imperfectly realized concepts, facilitating the transfer of representations of quantities to our mouths and, therefore, to the minds of others.[23]

4

BEYOND NUMBER WORDS:
OTHER KINDS OF NUMERIC LANGUAGE

Numbers govern our sentences. Even the words that I am writing now as I convey these thoughts can only be produced and comprehended in English with constant references to the quantity of items or concepts that are being talked about. Let me rewind and replay that sentence, highlighting the parts of it that refer to quantities: Even the word**s** that **I am** writing now as **I** convey **these** thought**s** can only be produced and comprehended in English with constant reference**s** to the quantit**y** of item**s** or concept**s** that **are** being talked about. There are no less than eleven places in that sentence alone where English grammar required me to distinguish the quantity of 'things' to which I was referring. The sentence is not an odd one in that respect. Furthermore, English is not unusual in its pervasive reference to quantities. Many languages require constant grammatical indications of the quantities being talked about, or of the amount of people involved in an interaction (e.g., 'I' or 'we'). In this chapter I hope to give you a sense of how important these numerical distinctions are to the world's languages. We shall see that grammatical number is very common, and that it also reflects a key point about our biology. But unlike number words, grammatical number says something about our neurobiology rather than about our hands.

This discussion begins with an overview of grammatical number, before offering some basic findings about human neurobiology that seem to motivate, at least in part, the patterns evident in the overview. In essence, the survey is a tour of numbers that are neither number words (like those discussed in Chapter 3) nor written numerals.

Number in Nouns

The place to start this tour is nominal number. Nominal number refers to inflections of a noun that signify the quantity of items denoted by the noun. In English, we have words whose form varies largely depending on their grammatical number. So if I am referring to one person, I say 'person', if more than one, I say 'people.' The number reference here is approximate in the latter case: we know only that there is more than one person, but without any precision beyond that. Other irregular noun pairs include 'tooth' vs. 'teeth,' 'mouse' vs. 'mice,' 'criterion' vs. 'criteria,' and other forms that give English-learners fits. More frustratingly to them, perhaps, are other irregular forms called 'zero' plurals—words that do not change whether there is one or more than one referent: 'deer' and 'sheep,' for instance. Other irregular plurals seem less idiosyncratic: We say 'children,' 'men,' and 'oxen,' yet such -en-suffixed words are not the standard form of plurality and have a different historical source than the most common form of nominal number in English. That common form is, simply, the addition of a sound to the end of a word that denotes more than one item. This sound varies a bit though it is generally written with s. Consider the following three word pairs in which the second word denotes more than one item, via a regular plural suffix:

(4.1)   cat   vs.   cats
         car   vs.   cars
         house   vs.   houses

In each case, the –s suffix signifies that more than one 'house,' 'cat,' or 'car' is being represented. In English, nouns that do not have any suffix are assumed to be singular, and nouns with an –s are understood to be plural. And you have been aware of this fact since you first began learning the language. Yet it is not quite that simple, even in a regular case in which –s is added to a word. If it is not clear why, read (4.1) again while focusing on pronunciation, and you can get a sense that the suffix used to signify a plural noun is not actually the same for all three words. In 'cats', the –s is what linguists call a *voiceless* sound, meaning your vocal cords do not vibrate as you produce it. In 'cars,' the sound is *voiced* and the –s suffix sounds like buzzing if you sustain it, because your vocal cords are vibrating. In 'houses,' the –s sound is again voiced, but a vowel is also inserted prior to its pronunciation, so we actually get a suffix that sounds something like '-uhz.' Nevertheless, despite this sound alternation, there is an underlying unity to the regular plural suffix in English, as evidenced by the fact that it is written with an *s* in all three cases.

Like English, many other languages have regular means of suffixing or prefixing nouns to signify the quantity of items being talked about. In most of these languages, a suffix or prefix is added to a word to denote plurality (i.e., that there is more than one of the elements represented by the noun). Consider the Portuguese equivalents of the word pairs in (4.1):

(4.2)   gato   vs.   gatos
         carro   vs.   carros
         casa   vs.   casas

In Portuguese as well, an *–s* suffix is used to represent plurality. Yet this does not mean grammatical number works the same in both languages. For one thing, the specific voicing alternation of English plural marking does not exist in Portuguese. For another, words adjacent to Portuguese nouns are also changed depending on plurality. For example, if I want to say 'my house' then I say *minha casa*, but to say 'my houses,' I must say *minhas casas*. In other words, the possessive pronoun *minhas* also inflects for plurality. Similarly, if I say 'the house' or 'the houses' in English, the article 'the' preceding the noun does not change. In contrast, in the Portuguese translations *a casa* and *as casas*, respectively, the article changes in accordance with the quantity represented by the following noun.

Nevertheless, the similarity of another European language may give the incorrect impression that the *–s* plural suffix is common to many or most languages. But as linguists have learned more and more about non-European languages around the world over the past few decades, they have found that the way nouns represent grammatical number can vary dramatically. Also, it has become clearer that there are many languages in which nouns do not change with the quantity of items they reference. Consider the following translations of 'cat' and 'cats', as well as 'house' and 'houses', in the Amazonian language Karitiâna:

(4.3) ombaky   vs.   ombaky
      'cat'   vs.   'cats'
      ambi   vs.   ambi
      'house'   vs.   'houses'

ly the nouns *ombaky* ('cat'—generally 'jaguar') and *ambi* ('house') do not change, regardless of how many cats or houses are being talked about. The same is true of all nouns in Karitiâna, since the language lacks nominal number (with the technical exception of pronouns like 'we').

In an incredibly comprehensive survey of 1,066 languages, linguist Matthew Dryer recently found that 98 of them are like Karitiâna and lack a grammatical means of marking nouns as being plural. So it is not particularly rare to find languages in which nouns do not show plurality. To speakers of European languages like English, in which marking nouns as singular or plural is a crucial skill required for fluency, it may seem odd that about 10 percent of the world's languages do not require speakers to do so. Yet what is more remarkable, I would maintain, is that such an overwhelming majority of the world's languages, about 90 percent of them, have a grammatical means through which speakers can convey whether they are talking about one or more than one thing. This strong tendency, evident in completely unrelated languages around the globe, suggests that the 'one / not one' distinction is crucial to us when we communicate. We take this fact for granted, but it is not clear a priori why this distinction should be so frequently referred to as we talk. To get a sense of why the distinction might be so prevalent in speech, it is useful to examine other categories of grammatical number that also exist around the world.[1]

Rather than simply separating nouns into those that represent one or more than one item, some systems of grammatical number also have another category referred to by linguists as 'dual.' This latter category is used when there are precisely two of the items being talked about. In Arabic, for instance, the suffix –an serves as a dual marker, and there is a separate suffix for plural nouns. While this may appear unusual to English speakers, it is interesting to note that Proto-Indo-European, the ancestor of English, also appears to have had a dual category. This is evidenced by the fact that Ancient Greek, Sanskrit, and other now-extinct descendants of Proto-Indo-European once used a dual category. In Ancient Greek, for instance, *o hippos* referred to 'the horse,' while *to hippo* referred to 'the two horses,' and *hoi hip-poi* meant, simply, 'the horses.'[2] In fact, Old English also had a dual category, and this category has weak vestiges in today's English. While our suffixes do not change form depending on whether there are two or more than two items being talked about, other English words reflect this distinction. If I say 'either of them,' as opposed to 'any of them,' you understand that I am talking about two people. Similarly, if I say 'both of them,' as opposed to 'all of them,' you also appreciate that there are two and only two people. So, apart from counting words like 'two' or 'three,' English can convey the distinction between one, two, or more than two items. In languages like Arabic, however, the dual category is much more regular and evident on suffixes frequently attached to nouns. Another modern language with dual suffixation is Slovenian.

Some languages indigenous to the Australian continent use a grammatical dual marker. Consider the following examples from Dyirbal, a language spoken on the Cape York Peninsula:

(4.4) Bayi Burbula miyandanyu
       "Burbula laughed."
(4.5) Bayi Burbula-gara miyandanyu
       "Burbula and another person laughed."
(4.6) Bayi Burbula-mangan miyandanyu.
       "Burbula and several other people laughed."[3]

These examples demonstrate that the *–gara* suffix is used when there are two people referenced, while the *–mangan* suffix is utilized when there are more than two being spoken about. The *–gara* suffix, however, is a dual marker. Technically it is an associated dual marker, since it is used to mean, in this case, "Burbula and another person," rather than "Two Burbulas." It is attached to personal nouns—unlike the English plural marker. Other Australian languages have dual markers that signify, more directly, that there are precisely two of the given noun being talked about. In Kayardild, the suffix *-yarrngka* serves such a function. For example, the word for 'sister' is *kularrin*, but if one says *kularrinjiyarrngka*, the word then means 'two sisters.'[4]

In languages that have a dual category, it is often more prevalent in pronouns or restricted to pronouns. (You may recall from grammar classes that pronouns are substitutes for other nouns, used mainly to refer to people talking or being talked about.) Consider the following pronouns from Upper Sorbian, a language spoken in a small region of eastern Germany:

(4.7)	ja	vs.	ty
	'I'	vs.	'you'
(4.8)	mój	vs.	wój
	'we two'	vs.	'you two'
(4.9)	my	vs.	wy
	'we'	vs.	'you all'[5]

The two pronouns in (4.7) are singular, referring to the first and second persons, respectively. The two pronouns in (4.9) are plural, again referring to the first and second persons. The middle example, (4.8), contains pronouns that cannot be translated into English without the word 'two,' but no number word is required in Upper Sorbian. That is because both pronouns in (4.8) are dual pronouns. While less common than plural inflections, dual inflections clearly do exist in contemporary languages and also are known to have existed in ancient languages.

Some languages have what linguists call *trial* inflections, which are used when there are exactly three of the items being referenced. However, the 'trial' category is apparently restricted to a small subset of the world's languages, more specifically some Austronesian ones. Consider the following sentence from Moluccan:

(4.10)   Duma   hima   aridu   na'a
          house   that   we three   own
          "We three own that house."[6]

The word *aridu* in this sentence refers exactly to three people. It is, therefore, a trial pronoun.

The list of categories of precise nominal number stops there, though. There are no clear cases of, for instance, 'quadral' grammatical number in the world's languages.[7] The only other major category of grammatical number that I have not mentioned is 'paucal.' This type of grammatical number is also uncommon but is used in some Austronesian languages to refer imprecisely to a few or more referents. Paucal number is used, for instance, by Boumaa Fijian speakers in a village of about sixty people. If a speaker in that village is communicating with a few or even over a dozen other people, she will use the paucal second person pronoun, *dou*. In contrast, if she is communicating with the entire village, she will use the plural second person, *omunuu*.[8]

In addition to varying in terms of their function, grammatical number categories can also vary extensively in terms of form. As we have seen in (4.1), (4.2), and (4.6), the form of the plural suffix is quite different in Dyirbal than in, say, English or Portuguese. Yet note that it is a suffix in all three languages, as the plural marker occurs at the end of a noun. This is not coincidental: suffixation is far and away the most common form that nominal number takes in the world's languages. Prefixation is not particularly rare though, and occurs in more than 10 percent of the 1,066 languages in the aforementioned worldwide survey of grammatical number. Here is an example of prefixed nominal number in Swahili:

(4.11)   ji-no   vs.   me-no
         'tooth'   'teeth'[9]

Nominal number may take more exotic forms besides markers
that are simply tacked on to the beginning or the end of a noun.
In some exotic cases grammars attach a plural marker to the middle
of a word, via a process known as *infixing*. Consider the following
word pair from Tuwali Ifugao, a language native to the Philippines:

(4.12)   babai   vs.   binabai
         'woman'   'women'[10]

Note that the infix *–in-* marker is added to the middle of *babai* to
make it plural.

Another exotic method used in some languages to pluralize a
noun is termed *reduplication* by linguists. In reduplication, one or
more syllables of a word are repeated to signify that a given noun
refers to more than one item. Tuwali Ifugao also utilizes reduplication, as in the following example, in which the first syllable of a
word is reduplicated to signify plurality:

(4.13)   tagu   vs.   tatagu
         'person'   'people'[11]

And the list of tricks that grammars use to turn singular nouns
into plural ones does not stop there. In the case of *suppletion,* for
example, a completely different word is used to express the pluralized version of a singular noun. The word for 'woman' in modern
Arabic is *mar'ah,* while the word for 'women' is *nisa.* In the Endo
language of Kenya, the word for 'goat' is *aráan,* while the word
for 'goats' is *no.* Suppletive plurals are difficult to learn, as each
plural / singular pair must be memorized. Other systems of gram-

matical number are even more onerous to acquire. Latin, Russian, and various other languages use completely different number suffixes depending on the case of the noun, for instance whether a noun is a subject or an object in a sentence.

We can draw a few conclusions from this brief survey of grammatical number in the world's nouns. First, as we saw in the preceding paragraphs, nominal number can vary extensively in form. In most languages it surfaces in suffixes, but in some languages it occurs in prefixes and, in a few, through more exotic forms like reduplications. We have also observed that some languages, like Karitiâna, do not use nominal number at all. Despite the many kinds of nominal number, the world's languages also exhibit some clear tendencies with respect to the function of this grammatical phenomenon. Most languages have singular and plural categories. Others have singular, dual, and plural. And finally, a few languages also have trial inflections. Yet, and this is a key point, no languages in the world have grammatical means of precisely referring to 4, or 5, or 6, or any other larger quantity—they must use number words to refer to such quantities. Clearly the world's grammars gravitate toward distinguishing between 1, 2, and 3 precisely, and all other quantities approximately. As we shall see below, there is a likely neurobiological basis for this gravitational pull.[12]

Number in Other Kinds of Words

While grammatical number is typically evident in nouns, since it usually indicates the number of people or other entities being talked about, languages may also change other parts of the sentence depending on the quantities being discussed. Languages commonly require some change to the verb of a sentence in accordance with how many things serve as the sentence's subject. This pattern is familiar to speakers of English and other European languages. Consider the following two pairs of English sentences:

(4.14)   The car is fast.   vs.   The cars are fast.
(4.15)   He runs slowly.   vs.   They run slowly.

In the first pair, we see that the English 'be' verb changes from *is* to *are*, depending on whether the subject is singular or plural. Linguists say this verbal change is used to show *agreement* with the grammatical number of the subject. In the second pair of sentences, we see that the verb is suffixed with *–s* when the subject is singular and receives no suffix when it is plural. Of course, in this pair the suffixing also conveys information about when the running happened. (I only say 'he runs' in the present.) In fact, verb suffixes often mix together grammatical number with some other category like tense. Languages are messy.

Consider two more sentence pairs illustrating grammatical number through verbal agreement, taken from Portuguese:

(4.16)   Ele foi ontem.   Eles foram ontem.
          "He went yesterday."   "They went yesterday."
(4.17)   Marta jogou futebol.   As mulheres jogaram futebol.
          "Marta played soccer."   "The women played soccer."

The verb *foi* ('went') changes to *foram* in (4.16) depending on how many people went the day before. In (4.17) the suffix on the verb *jogou* ('played') is altered in accordance with whether one or more than one person is playing. Sentences (4.14–4.17) reflect a strategy commonly used by the world's languages, wherein the verb changes somehow when a plural noun is the subject of the sentence. In some languages, this strategy is modified, and the verb 'agrees' with an object instead of the subject. This is evident in the European isolate language Basque, which is called an *isolate* because it is unrelated to any other known languages:

(4.18) Nik   luburuak irakurri di-tut
        I   books   read   3rd person plural-have
        "I have read the books."[13]

In this case, the auxiliary verb *tut* ('have') is prefixed with *di-*, which implies there is more than one book being talked about, as opposed to more than one person doing the reading.

Already we have gotten some sense that grammatical number is ubiquitous in the world's languages, but also that it takes many different shapes. It can be denoted simply with a plural suffix attached to a noun, or with a dual pronoun used to refer to two people, or with a prefix attached to a verb that agrees with the number of a noun somewhere else in the sentence, or with other changes to nouns and verbs.

And languages do not stop there. Consider the indefinite articles in English. I say "a car" and "a computer," but it is clearly ungrammatical to say "a cars" or "a computers." While the indefinite article *a* is not a number word, it obviously conveys some quantitative information. Many other languages share this feature, and speakers use different articles in accordance with the quantity of given referents. In German, for instance, I could say *das Auto* for 'the car' but must change the definite article to *die* if the noun is pluralized to *Autos*. Or consider English demonstratives like 'this' and 'that,' which highlight a particular entity being spoken of while also communicating something about how close the entity is to the speaker. I can say "this pen here" or "that pen over there," but if I am talking about more than one pen here or there, I need to change the demonstratives: "these pens here" and "those pens over there."

Some languages have words with a vaguely similar feel and function to demonstratives, referred to as *classifiers* by linguists. Classifiers are words or parts of words that categorize nouns that occur next to them. They do not categorize nouns according to distance, like demonstratives, but typically according to some feature like the animacy or the function of the item being referred to by the noun. Interestingly, classifiers are often attached to number words when things are being counted. Consider the following examples from Yagua, a language indigenous to northwest Amazonia:

(4.19)	tï-kïï	varturu
	one-classifier	woman (married)
	'one married woman'	
(4.20)	tïn-see	vaada
	one-classifier	egg
	'one egg'[14]	

The classifier suffix attached to the number 'one' varies depending on whether a person or an egg is being talked about. Many languages have classifiers that surface during counting, including two of the most widely spoken languages in the world—Mandarin and Japanese. In some Mayan languages, nouns are grouped into dozens of categories that are evident during counting. English has some hints of a classifier system. When we count so-called mass nouns like 'sand,' 'dirt,' and 'clay,' we must categorize their shape. I cannot grammatically say 'thirty clays,' or 'thirty dirts,' or 'thirty sands.' I need to add words like 'clumps of,' 'mounds of,' and 'grains of,' respectively, while also changing the nouns to their singular forms, to make such phrases grammatical. In contrast, *count* nouns like 'cars,' 'pencils,' and 'books' do not need such help. 'Thirty cars' makes perfect sense; 'thirty clumps of car,' not so much.

Clearly grammars have myriad ways to distinguish the quantities of the referents being talked about. But note that all these ways, including grammatical phenomena like number verb agreement and singular definite articles, are devoted to dividing small quantities, particularly 1 and to a lesser extent 2 and 3, from other quantities. Grammatical number is only approximate, however, when it comes to higher quantities. This approximating tendency is also evident in actual words for quantities. In Chapter 3 we focused on words for exact quantities, but it is worth noting here that languages also have loose-fitting numberlike words as well. English has such words: *a few, a couple, many, several,* and so on. Such terms likely exist in all languages, so potential examples are pretty much limitless. For instance, Yucatec Maya has words such as *yá'ab'*, meaning 'many' or 'much.'[15] Surprisingly, perhaps, some languages rely exclusively or almost exclusively on such approximate number words when describing quantities. We will consider such languages in Chapter 5 when discussing anumeric people.

Other imprecise number words may be used to indicate that more than one of a specific kind of item is being talked about. So, if I am referring to a group of many hooved animals, I can use *herd.* If I am talking about a similarly numbered group of swimming animals, I can refer to a *school* of fish, or perhaps a *pod* of dolphins. There are dozens of such words for groups of animals in English. A *gaggle* refers to a group of geese that are not flying. If they are flying, the most appropriate term is *skein,* at least if one cares what pedantic sorts think. If I were talking about ducks instead of geese, I would say a *flock.* Many English speakers are unaware of such distinctions, and understandably so, given their limited usefulness. Yet the distinctions are nevertheless available, hinting at another way in which languages emphasize the distinction between one and more than one.

This distinction also surfaces in quirky variations between verbs. For instance, if I observe one elephant moving quickly I might say it is running. If there is a large quantity of elephants moving in the same manner, I might say they are *stampeding*. This verb change is due to the number of elephants, of course, not the number of times that a given elephant ran. Intriguingly, some languages use such verbal differences to refer to the number of times an event has occurred, rather than to the number of entities involved in the event. This phenomenon of *pluractionality* is evident in the Chadic language Hausa of the African Sahel. For example, in that language the verb *aikee* means 'to send,' as does *a"aikee*. The *a"*- prefix on the second version of the verb signifies that something was sent over and over. So the verb changes in accordance with the number of actions of sending, not the number of people sending or being sent something, nor the number of objects being sent.

The Amazonian language Karitiâna has special verbs whose meaning implies that multiple items are involved in an event. This is somewhat surprising since, as I mentioned earlier in this chapter, Karitiâna does not have nominal number (apart from pronoun distinctions). What it does have are a few verbs that have an inherently plural feel. For instance, the verb *ymbykyt* means "several people arrive." The verb *piit* means "take a bunch of things." This verb is used regardless of how many people are doing the taking. Other 'plural' verbs are used when describing actions such as 'running,' 'going,' and 'flying.' In a brief study carried out with two dozen Karitiâna speakers, I found evidence that the use of distinct plural verbs impacts how the speakers think of certain actions, when contrasted to speakers of languages like English that do not have such alterations.[16]

Grammatical number is often thought of as a simple distinction between singular and plural nouns. We have seen that it is much more than that. Grammatical number is nearly omnipresent in the world's languages, but it is a shape-shifting beast. In many languages it does indeed take the shape of suffixes referring to quantity distinctions. But sometimes these distinctions are not as simple as 'one' vs. 'many.' The distinctions can break apart the number line in more specific ways, perhaps 'one' vs. 'two' vs. 'many.' Furthermore, we have seen that grammatical number does not simply surface in nouns. Verbs may also indicate the quantity of referents being talked about or may reflect the amount of actions being described. Other words, like articles and classifiers, also reflect the human tendency to refer to quantities, even when those quantities may seem wholly irrelevant to a particular conversation. Languages do not simply reflect this tendency, either. They also reinforce our numerical fixation by requiring continuous reference to quantities.

Despite the varied nature of grammatical number, we have also seen some remarkably strong tendencies in the grammars of the world's languages with respect to this phenomenon. First, the vast majority of languages have grammatical number. It is one of the most prevalent features in the landscape of the world's grammars, which is increasingly recognized as being dramatically diverse. Second, and just as crucially, while grammatical number strategies take disparate forms across the world's languages, their functions are remarkably similar. First and foremost, grammars tend to group quantities into one of two categories: 1 or everything but 1. In cases where grammars appeal to more nuanced categories, they are still remarkably restricted. Grammars refer to, at the most, three exact quantities—1, 2, and 3. No languages have, for instance, suffixes that refer precisely to 5 or 10, despite the prevalence of quinary and decimal patterns in the world's number words. Given the range of esoteric meanings referred to by all sorts prefixes and suffixes in the world's languages, the limited range of quantities referred to by grammatical number is actually fairly remarkable. All of this begs an obvious question: why are human grammars so focused on quantities, but only in a fuzzy way—unless the quantities are 1, 2, or 3? If we want to be precise when referring to quantities greater than 3, we need to use number words instead of grammatical number. It is almost as though distinguishing larger quantities in precise ways does not come naturally to us, but distinguishing 1, 2, and 3 does. As it turns out, this is true. To understand why certain quantity distinctions are more natural to us we need to look at the cerebral equipment we use to understand quantities.

The Neurobiological Basis of Grammatical Number

The intraparietal sulcus (IPS) is one of the many valleys in the human brain. It stretches horizontally, in the parietal lobe, from the central portion of the cortex toward the rear. The IPS is a hotbed of numerical thought, a point we explore more fully in Chapter 8. One of the remarkable things about the numerical thought that happens there is that some of it is both ontogenetically and phylogenetically primitive. In other words, some of that thought happens really early in our individual development (ontogeny) and also seems to be ancient in our species and in related ones (phylogeny). To some extent, humans and other related animals are wired for numerical thought.

To *some* extent. And the "some" here is one of key points underscored by this book: the numerical tools that our brains offer us, apart from culture, are pretty blunt. But they certainly exist. One of the distinguishing features of the native numerical processing that takes place in the IPS, governed by our innate neurobiology rather than by cultural convention, seems to motivate the grammatical patterns evident in the preceding sections. Because it turns out that humans are innately predisposed to differentiate smaller quantities, in particular 1, 2, and 3, from one another and from larger quantities. Distinguishing, say, 1 from every other quantity, comes naturally to us.

A vast body of scientific literature in cognitive psychology, neuroscience, and related fields has demonstrated that humans are able to quickly discriminate small quantities of objects prior to any sort of mathematical training. (This point is made clearer in our discussion of infant cognition in Chapter 6.) This object-tracking ability is made possible by basic neurobiological features like the IPS. The ability enables us to accurately and quickly discriminate sets of 1, 2, or 3 items. When it comes to larger quantities, however, our innate neural mechanisms only offer us a fuzzy means of quantity differentiation. Consider this dramatic case as an illustration of how powerful our natural discrimination of smaller quantities is. Say you are walking down an alleyway in New York City, and you witness a small group of anonymously dressed criminals standing over someone they have just attacked. Even if you have only a second or less to visually process the scene, if there are three or fewer attackers you will immediately recognize how many criminals are present. If you were later interrogated by police officers, you could confidently tell them how many assailants you saw, assuming that they were visually distinguishable. In contrast, say you walked down the same alley but instead saw six criminals standing over someone. If you only had a split-second to process the scene before they (or you) dispersed, would you be able to accurately and confidently tell how many perpetrators there were? Not really. When people are asked to quickly appreciate a given quantity of items, like visually distinguishable people, they can only approximate that quantity if it exceeds 3.

For quantities greater than 3, we are better at discriminating sets that differ in pronounced ways. So if the officer asks you if there were 6 or 7 assailants, you might give the wrong answer since the difference between 6 and 7 is not pronounced. However, if the officer happens to know it was either a gang of 6 or of 12 assailants, and gives you the choice between 6 and 12, you will choose the correct answer, because 12 is 100 percent greater than 6. But to be precise when perceiving quantities like 6, we need to count the items in question. A police report based on your eyewitness testimony would be quite accurate in the first alleyway scenario, but in the second scenario the testimony would be dubious. Maybe you did see six attackers. But on second thought perhaps it was seven. Or five. Or maybe, just maybe, it was eight. If you did not have the opportunity to count them, to use those verbal symbols for quantities we call number words, your account would be unreliable. The genetically hardwired mechanisms of the IPS and other cortical regions are only capable of exactly discriminating smaller quantities, as evidenced by experimental and brain imaging studies.

In some sense, this point may seem obvious—of course it is easier to distinguish smaller quantities. But the claim here, based on extensive studies by many researchers, is not simply that. Humans are not just a little better at distinguishing smaller quantities, and our mathematical fuzziness does not just increase gradually in accordance with the number of items being perceived. Instead there is a clear gap between how we think of 1, 2, and 3 when contrasted with all other quantities. Framed another way, we are natively predisposed to think of these quantities in exact ways and of all other quantities in approximate ways. As renowned psychologist Stanislas Dehaene notes, we have a "number sense" that helps us exactly recognize some quantities. More precisely our brains, especially the IPS in our brains, house two number senses: an exact number sense and an approximate number sense. In reality the former sense is an object-tracking capacity, often referred to as the "parallel individuation" system, so it does more than allow us to think quantitatively. Yet one of its features is that it enables us to precisely recognize a group of small quantities—like a group of 1, 2, or 3 attackers. For that reason, I refer to it as an "exact number sense," a term that serves a mnemonically useful contrast with the "approximate number sense." The latter sense enables us to estimate the amounts of larger quantities, like a group of 5–7 attackers. These senses, discussed further in subsequent chapters, are the building blocks to the distinct quantitative reasoning associated with our species. But they come nowhere near explicating full mathematical thought.[17]

This background on our basic neurobiology casts new light, it would seem, on the discussion we have been having about grammatical number. First, the innate numerical neurobiology of our species likely motivates one of the key findings from our survey: grammatical number is everywhere. It is found in the vast majority of languages and, in the languages in which it does exist, it generally has prominent effects on how people produce sentences and individual words. It can surface in nouns, verbs, articles, classifiers, and other word types. Second, when grammatical number refers to specific quantities, these quantities are severely restricted. Grammatical number tends to distinguish 1 from other quantities, but 2 and 3 can also be specifically referred to. Third, grammatical number often makes reference to large quantities, but always in an approximate way.

In Chapter 3 we saw that most number systems are motivated by human biology. In the great majority of languages, numbers show clear historical connections to the fingers and hands. The findings discussed in this chapter demonstrate that grammatical number too has common features across the world's languages. These common features are not due to characteristics of human appendages, however, but apparently stem from features of the human brain. Our grammars refer to numerical concepts that, not coincidentally, are easily and natively filtered by our brains—especially the IPS.

If these suggested motivations for number words and grammatical number are accurate, we might expect that number words for very small quantities would differ in some manner, when contrasted to number words for larger quantities. This prediction is generally supported by the linguistic evidence. Number words for quantities like 1, 2, or 3 tend to have sources that are lost to time, and origins that are clearly distinct from larger numbers in the same language. Unlike numbers for quantities like 5 or 10, humans do not typically name lower quantities after body parts. (As with most generalizations about human languages, though, there are some rare exceptions.) And we do not need to. We can simply make sense of such quantities with our minds, without needing a physical proxy to help us digest the quantities. We do not need to refer to matching quantities in our mind-external world to invent words for 1, 2, or 3.

Smaller numbers differ from larger ones in another crucial way, one that also supports the suggestion that they are based directly on innate concepts. As researchers like aforementioned psychologist Stanislas Dehaene have pointed out, small numbers are used much more frequently than other numbers. In fact, in written language the numerals 1, 2, and 3 are more than twice as common as all other numerals. Much of the reason for this frequency is that such numbers are easier to conceptualize accurately and quickly. It is not the case, for instance, that nature presents things more frequently in packages of 1, 2, or 3, at least not to the degree evident in language. Furthermore, the frequency of written and spoken numbers, in all languages in which number-frequency has been studied, does not decline in a regular manner as the number increases. Instead, there tends to be a steep drop-off in the rate of occurrence of the numeral 4, when contrasted to the numeral 3. Additionally, in English and other decimal languages, numbers like 10 and 20 are also incredibly frequent, much more so than other large numbers. This suggests that the greater frequency of some numbers in speech and in texts is not because some quantities are simply more common in the world around us. Instead, some numbers are more frequent because they are more easily filtered by our brains and our bodies.[18]

The ease with which humans process smaller quantities is reflected in the patterns we have described vis-à-vis grammatical number. But it is also evident in the frequency with which smaller numbers are used; in the unclear origins of smaller numbers; and, it should be mentioned, in ordinal numbers. Ordinal numbers indicate the position of some item or event in a sequence. I can say, for instance, that "Germany is the third country to win the World Cup a fourth time." And, in writing that sentence exemplifying ordinal numbers, I have illustrated another way in which languages distinguish quantities 1, 2, and 3 from others. Consider the list of ordinal numbers in English: *first, second, third, fourth, fifth, sixth, seventh, eighth, ninth, tenth, eleventh, twelfth,* and so on. Notice how the first three words in this set have irregular endings, while all remaining ordinal numbers end in *–th.* Yet again, we see that numbers associated with lower quantities are given special linguistic status, reflecting indirectly their unique status in our brains.

As a final example of the way in which smaller numbers are treated distinctly in human language, consider Roman numerals. These numerals evolved from a tally system based on linear marks. In Roman numerals, smaller quantities are represented simply via lines: I (1), II (2), and III (3). Yet larger quantities are treated differently, because, unlike a series of three lines, larger series are cognitively unwieldy: VI is easier to discriminate than IIIIII. The latter series of lines is difficult to precisely quantify, when contrasted to, I, II, or III, which can be immediately subitized. This facilitated differentiation of 1–3 is even evident in the representation of 4 in

Roman numerals: IV. Smaller quantities, specifically 1, 2, and 3, are treated distinctly and more directly, when contrasted with all other quantities.[19]

The simple symbols for smaller quantities in Roman numerals is just another example of the patterns observed worldwide in systems of grammatical number and in words for small quantities. Human neurophysiology enables us naturally to think and talk about lower quantities. Alternate explanations of the patterns in question are problematic. There is no convincing evidence, for example, that small quantities are somehow more prevalent in the natural environments of humans. There is, however, evidence that our innate number sense makes us better discriminators of smaller quantities.

Scanning the world's languages, one gets a sense that grammatical number is exuberantly diverse. This impression is not wholly inaccurate, since grammatical number does come to life in diverse forms. But more careful inspection reveals some commonalities in the functions served by grammatical number. These commonalities are likely due to our basic neurobiology—much as common patterns in the number words of the world's languages are due to basic facts about our biology.

Conclusion

Still not 2 years of age, my son was sitting in the backseat of our car as we drove over the Rickenbacker Causeway—a bridge connecting the island of Virginia Key to the city of Miami, splitting the turquoise Biscayne Bay. As our car climbed the bridge, my son looked out the window to his right, gaped at the bay extending to the ocean on the horizon, and yelled "water!" He then pivoted his head to the left, focused now on the portion of the bay that stretched to the city's shoreline. Taken aback by the existence of another seemingly separate body of water, he exclaimed "Two waters!" Two waters. And why not? For a child that has not yet grasped the distinction between count and mass nouns, such an expression makes perfect sense. But my son's vivacious declaration is telling in more fundamental ways. From a really early age, often before the age of 2, speakers of English and most other languages realize that they must pluralize nouns if those nouns refer to more than one item. Kids are clearly capable of learning this distinction early in life. Languages place a lot of importance on grammatical number, an importance our brains handle adroitly, because they are predisposed to do so. This predisposition likely motivates the worldwide pervasiveness of grammatical number.

The ubiquity of grammatical number is remarkable, particularly given our growing awareness of the diversity of human speech. In the past several decades, as linguists have ventured into the remote corners of the world while documenting unrelated languages, we have learned that languages can be dramatically unalike. Many linguists now consider the diversity of languages to be the most notable feature of human communication. Some languages lack tense, others do not have distinctions between colors like red and yellow, others lack clear grammatical subjects. Some languages have as few as ten meaningful sounds, while others have more than one hundred. And so on. Languages are radically diverse, and some language researchers now believe that this diversity reflects, in small part, the capacity of languages to gradually adapt to different environments. (My own research with colleagues suggests that some aspects of linguistic sound systems evolve in ways that are subtly influenced by environmental factors, such as extreme aridity.)[20]

Despite such remarkable diversity and adaptability, there are extremely strong tendencies evident in the world's languages with respect to how they treat quantities. We have seen in this chapter that grammatical number exists in the great majority of the world's languages. Grammars are number obsessed, and their obsession is often focused on differentiating a few small precise quantities from fuzzy larger ones. In this way, grammatical number mirrors in function our cerebral architecture, which comes pre-equipped to exactly differentiate only smaller quantities.

Despite the common nature of grammatical number, though, it is also the case that some languages do not have it. Similarly, in Chapter 3 I noted that not all languages have number words either. In Chapter 5 we consider a pivotal issue in our quest to understand the role of numbers in the human story: what happens when people do not utilize grammatical number, number words, or any other symbolic representations of quantities? Next we take a look at worlds without numbers.

PART 2

Worlds without Numbers

5

ANUMERIC PEOPLE TODAY

Occasionally as a child I would wake up in the jungle to the ca-cophony of people sharing their dreams with one another—impromptu monologues followed by spurts of intense feedback. The people in question, a fascinating (to me anyhow) group known as the Pirahã, are known to wake up and speak to their immediate neighbors at all hours of the night. At times this practice can ir-ritate outsiders trying hard to sleep. To a young boy like I was, though, the voices echoing through my family's large hut allayed my nighttime jungle-related fears, comforting my psyche even as I understood little of what was being said. After all, the voices sug-gested the people in the village were relaxed and completely un-concerned with my own preoccupations. Their midnight voices seemed devoid of the particular worries that sometimes kept me still in my hammock when awakened by some unrecognizable jungle noise. When awakened by the dream sharers, I usually drifted back to sleep in little time.

The Pirahã village my family lived in was reachable via a one-week sinuous trip along a series of Amazonian tributaries, or alternatively by a one-hour flight in a Cessna single-engine air-craft. That flight ended with the pilot somehow settling the plane's wheels on a narrow grass strip that, prior to the plane's approach, was invisible amid the surrounding sea of trees. Even today the Pirahã remain isolated, their culture largely unchanged since my childhood, or for that matter since their first contact with Brazilians more than two centuries ago. They still live in small riverine settlements with physically modest dwellings, occasionally dispensing with dwellings altogether and sleeping on wood slats along the white river beaches that crop up in the dry season. And they still wake up in the middle of the night, seemingly in mid-conversation, as they share their experiences from their dreams.

Along with my parents and two older sisters, I spent many months of my childhood with this small group of hunter-gatherers in the heart of Amazonia. This people is remarkable for numerous reasons and, occasional night fears aside, my childhood memories of them are pleasant, bordering on idyllic. My parents took us to spend time with them as part of their own work at the time, as evangelical Bible translators. But the Bible is not the only outside artifact we brought with us as a family. Other imports, some would say less objectionable ones and seemingly more palatable ones to the Pirahã, included Western medicines that saved many children's lives. The importation of other foreign items was generally less successful. For instance, most outside food products puzzled the Pirahã as much as their habit of eating one another's lice puzzled me and my sisters. A Pirahã man once asked why we felt the need to spread a bloodlike substance, known to us as ketchup, all over our food. Similarly, while we were once eating salad, another man pointed out to other Pirahã onlookers that, yet again, we were oddly eating leaves. Another category of unsuccessful imports was Western symbols of various kinds. This included alphabetic letters, since unlike most indigenous groups the Pirahã proved uninterested in writing their own language. And these unsuccessful imports also included numbers, which were rejected wholesale by the people.[1]

On many of those nights in the village, prior to heading off to "bed" in our hammocks, my parents held math lessons for the Pirahã, increasing participation levels by offering one well-regarded good from the outside: popcorn. By the insect-attracting incandescence of gas lamps scattered through our family's hut, open to the elements along the banks of the black-water Maici River, they attempted to teach the people math in their native language. The attempts were unsuccessful, for a variety of reasons, it turns out. Perhaps the most prominent of these reasons is simply that the Pirahã language lacks all precise numbers. Typically, when one culture adopts another's number system, the adopting culture is at least aware of what number words are. In the case of the Pirahã, however, numbers were entirely foreign. Not just the specific Portuguese number words my parents sought to teach them, but also the existence of precise number words and, crucially, even the recognition of most of the exact quantities they denoted.

As a young boy, the difficulty that Pirahã adults faced when trying to learn numbers puzzled me greatly. In large measure it was puzzling because it was clear, even through my young apperception, that these were not people with a learning impairment of some kind. There is no prevalent genetic abnormality among the Pirahã that explains the challenges they faced then, and face now, when it comes to learning numbers. Furthermore, the few Pirahã who have been raised in outside cultures exhibit no such difficulty. In fact, in many other respects I marveled at the cognitive dexterity of the people. Some of this marveling was no doubt due to my own youth, but the experience that motivated it was not trivial: I (like my sisters) played with their children, following them through the jungle—at times I would have been literally lost without them. I watched them fish better than I could, recognize fruit sources in ways I could not, and generally felt they were unmatchable when it came to many mental skills that were so crucial in their ecology.

Yet here, at the occasional lamp-lit math lesson, I was the one on superior footing even when compared to the adults.

These people are not unique in having linguistic and cultural barriers placed between themselves and the acquisition of basic mathematics. In fact, hundreds of kilometers to the east is a group that has faced analogous struggles: the Munduruku. The Munduruku are a large and formerly warlike tribe on the upper Tapajos River, a principal tributary of the Amazon. They took to tapping rubber during the late nineteenth century—and some still do so today. Working hard, they gathered large quantities of rubber, particularly during the height of the Amazonian rubber boom in the early part of the last century. As historian John Hemming notes, though, "they were woefully uncommercial, easily duped because they did not understand arithmetic. Regatão (river boat) traders sold them goods at a fourfold mark-up, including quantities of cachaça rum and useless patent medicines. They were of course paid poorly for their rubber."[2]

As we saw in Chapters 3 and 4, languages vary extensively in the way that they encode numerical concepts. Some languages have number systems that allow for the generation of an unlimited number of terms for quantities. Many languages have less robust systems, however. The Pirahã and Munduruku languages certainly fall into the latter category, and Pirahã in particular may represent the most extreme case: a spoken language without any precise number terms, not even for 'one.' And this claim is not just anecdotally based, either. The numberless nature of the language was first brought to the attention of the scholarly community by my father, Daniel Everett, who eventually traded in his missionary career for one of a researcher.[3] His comments on the matter resulted in several psychologists and others conducting experiments to test whether the language was indeed anumeric. For instance, let us consider one task conducted by psycholinguists about 10 years ago.

In the task, the Pirahã were presented with an array of objects—spools of thread. They were then asked to name the quantity contained in the array. All fourteen subjects who participated in the experiment used the word *hói* to refer to one object. This term is best translated as 'small size or amount' and is the lowest quantity-related term in the language. The next smallest quantity-related term is *hoí,* which differs from *hói* only according to vowel tone. This tonal distinction alters the meaning of the word, with the meaning changing to, roughly, 'a couple or a few.' But note that the meaning of *hói* bleeds into that of *hoí,* as evidenced in the experimental results of the study in question. The researchers found that, unlike those cases in which the fourteen participants were asked to label one spool of thread, there were some discrepancies when they were asked to label higher quantities. For two spools of thread, *hoí* was used in most cases, but, remarkably, some speakers utilized *hói* (demonstrating that the latter word is not simply 'one'). For three, the results were also mixed. In fact, as the quantity increased, the usage of *hoí,* decreased, but the transition from *hói* to *hoí,* and subsequently from *hoí* to *baágiso* (the word meaning, roughly, 'many' or, literally, 'bring together'), was gradual. In contrast, speakers of English performing the same task only use *one* for 1 thing, *two* for 2 things, *three* for 3 things, and so on.[4]

This is a remarkable finding, since supported by other experimental work. It implies that the three number-like words in Pirahã are not in fact precise numbers. These terms are used only in an approximate sense, much as we use phrases like 'a few' or 'a couple of' in English. In contrast to speakers of English, however, speakers of Pirahã have no precise alternatives, such as 'three' or 'two.' The Pirahã language is anumeric—it even lacks the sorts of grammatical number distinctions discussed in Chapter 4. This is an incredible feature of the language and of Pirahã culture more generally: they have chosen not to incorporate precise numbers into their daily experience. And, while the Munduruku do have number words, most of these are also used in imprecise ways. In a landmark study published in *Science* in 2004, a team of cognitive scientists demonstrated that most of the number words in that language also have fuzzy meanings.[5]

The motivation for the cultural obstacles to the incorporation of precise numbers among such populations is debatable, but the rigidity of the obstacles is unusual, since number systems generally flow across cultures, particularly after prolonged contact with outsiders, such as that experienced by these two indigenous groups. Put differently, when cultures confront other groups with more words for quantities, they usually borrow some or all of those words, or at least the concepts conveyed by those words. This borrowing is understandable, since numbers are so useful. Given this oft-observed tendency, it is perhaps surprising to find that some cultures have not adopted the more extensive number systems of groups they have encountered, regardless of how difficult this adoption may be in their particular cases. Yet the Pirahã, the Munduruku, and some other cultures have remained anumeric or at least largely anumeric. (There are signs this situation is now changing in both groups, however.) As we will see next, this choice has pervasive effects.

Looking for Answers in the Jungle

In a highly publicized study also published in *Science* in 2004, a psychologist from the University of Pittsburgh demonstrated experimentally that the absence of number words in the Pirahã has a profound effect on the quantity discrimination abilities of the members of that culture. The psychologist, Peter Gordon, conducted a series of experimental tasks over the course of two summers visiting the Pirahã, with the assistance of my parents. The results of

5.1. A Pirahã family in a typical canoe on a tributary of the Maici River. Their clothing is one of the few imports from the outside world. This photograph was taken in 2015 by the author.

these tasks demonstrated in a clear and replicable fashion something there had been stories of for some time: the Pirahã struggle with the precise differentiation of quantities above three. In many respects the experiments conducted by Gordon were similar to tasks my parents had attempted in their aforementioned lessons with the people in the early 1980s.[6]

In the flurry of attention directed at the Pirahã subsequent to the publication of Gordon's study, the depiction of these people was often less than accurate and sometimes grotesquely distorted. To some anyway, they came to represent a sort of atavistic way of life—a throwback to stone-age days, a relic of an anumeric time. Others suggested that the people's difficulty with mathematical concepts may owe itself to inbreeding, some sort of recessive gene(s) expressed due to population bottlenecking. Both accounts miss the mark entirely, of course. The most plausible conclusion derivable from Gordon's study is, simply, that the Pirahã are a group of hunter-gatherers who have chosen not to utilize numbers and who, as a result, do not wield the cognitive advantages afforded by this tool. To better understand this interpretation, it is worth examining the findings of Gordon's study and subsequent work by others, including myself.

But first, some background on how humans generally perceive numbers. As noted in Chapter 4, human beings are innately equipped with two mathematical "senses." These aptitudes or genetic endowments would hardly be considered mathematical in nature by most adults, given their simplicity. But they are foundational to the enterprise of thinking with numbers. First, there is our approximate number sense, which is our innate ability to estimate quantities. Human infants seem to be born with this system, and it enables them to recognize large differences between quantities. As we shall see in Chapter 6, newborn babies can distinguish, for instance, eight items from sixteen items. They are able to do fuzzy math with larger quantities. The second major innate mathematical capacity humans have is the ability to exactly differentiate quantities of three or fewer. In other words, members of all human populations, at all ages, can differentiate one object from two objects, two from three, and three from one. In this book I refer to this capacity as the exact number sense, for convenience of contrast with the approximate number sense. (Since it generally serves to keep track of small sets of objects in parallel, it is usually referred to as the "object-file system" or the "parallel individuation system" in the field of psychology.) As I mentioned in Chapter 4, these two senses are housed largely in the intraparietal sulcus, or IPS, of our brains.

The existence of these different primal mathematical abilities has been well established by now. But an awareness of these abilities still leaves open tantalizing questions: How are humans, and not other species, able to unite these abilities? How do we tie these abilities together? What enables us to transfer exact recognition of small quantities to the larger quantities handled more naturally by the approximate number sense? Broadly speaking, there are two basic answers to these similar questions. The first answer is a nativist one, according to which these innate abilities are fused in the human brain simply because that is how the human brain operates. According to such a perspective, we are genetically endowed not just with the approximate and exact number senses. We are also endowed with the capacity for gradually tying these abilities together somehow. In other words, one of our distinguishing features as a species is simply that our brains are wired for numbers, and that as we develop we naturally realize that quantities like 5 and 6 differ from each other. (Even if labels like *five* and *six* may facilitate this realization, the realization is possible in their absence.) The second potential answer is a culturally oriented one: humans learn to unite their innate mathematical capacities only if they are exposed to numbers after being embedded in a numerate culture while speaking a numeric language. Such an account suggests that humans learn to differentiate quantities greater than 3 in precise ways only when they learn culturally shared symbols for precise quantities—when they learn number words.

These two potential accounts of the origins of distinctly human numerical cognition make largely similar predictions. Both accounts predict that, as humans develop in numerate societies, they will gain an appreciation of the exact differences between quantities beyond 3. Since nearly all the world's cultures are numerate, for some time it was challenging to offer unambiguous support for one or the other hypothesis. A proponent of the culturally oriented account might suggest, for example, that kids only learn to truly differentiate quantities greater than 3 when they learn to count. A proponent of the nativist position might counter, though, that we learn to count when our brains are developed sufficiently for the task. One way to shed light on the issue would be to offer data from a healthy group of adults who live in an anumeric society. If a population lacked number words or other forms of numeric culture, would its members learn to differentiate most quantities precisely? Or would they be restricted to the simpler approximation offered by our cerebral hardware? An affirmative answer to the former question would offer strong support for a nativist position, while an affirmative answer to the latter question would offer clear support for a culture-based account.

With such issues in mind, the relevant studies carried out among the Pirahã have focused on illuminating the capacity, or lack thereof, that these people have for exactly differentiating quantities greater than 3. Gordon's study consisted of a series of quantity-recognition tasks carried out with adults in two villages. The point of the tasks was to get at this central question: Do healthy anumeric adults precisely and consistently distinguish quantities greater than 3 from each other? Can they consistently discriminate 6 items from 7 items, or 8 from 9, or even 5 from 4? If they cannot, this would suggest that familiarity with counting and number words is essential not just for math but also for the mere recognition of most quantity distinctions.

Gordon's experimental tasks fell into two broad categories. For one category of task, the experiment's participants were asked to match the number of objects presented on a surface in front of them by placing the same number of objects on that surface. For this sort of task, the objects used were small AA batteries. There were several variants of the tasks with the AA batteries. In the most basic "line matching" task, the participants were simply presented with an evenly spaced line of AA batteries. The batteries were placed close to one another so as to be perceived as a group, but a group comprising distinguishable items. The participants were then asked to place the same quantity of AA batteries in an array parallel to the original line, after the task was first modeled for them accurately. (This is true of all the quantitative tasks since conducted among this group—experiments are always modeled by the researchers first, to reduce any confusion.) In an "orthogonal matching" task, the participants were again tasked with making a line out of AA batteries. The line was supposed to be equal in number to the original line presented to them, but in this case was rotated ninety degrees. (Rather than parallel to the original line as in the basic line matching task.) In another related "brief presentation" task, the subjects were presented with a line of batteries and then asked to produce a line of the same quantity, but only after the original line was hidden from view. In the second general category of task utilized by Gordon, the participants were asked to demonstrate recognition of the quantity of items in a container. So, for example, Gordon placed nuts in an opaque can, one at a time, in front of the participants. (The participants could see the sides of the can from their perspective, but not into the can.) He then removed those same nuts, one at a time, still in full view of the subjects. After each nut was removed, the Pirahã participants were asked to signal whether any nuts remained in the can.

As Gordon conducted his experiments, a clear pattern emerged in the Pirahã responses, regardless of the particular demands of a given task. In short, the Pirahã struggled markedly with precisely differentiating quantities. Crucially, though, this struggle was really only apparent for quantities exceeding three. For instance, when participants were asked to simply match a line of batteries with another line, the quantity of the second line always equaled the first when 1, 2, or 3 batteries were first presented. When the initial line included 4 or more batteries, however, errors began to surface in the responses. And the ratio of errors increased in direct proportion to the number of batteries presented. A similar pattern was observed for the orthogonal task as well as the brief presentation task, though in the latter cases more errors were observed—unsurprisingly, given the more difficult nature of the tasks. For the nuts-in-a-can test, the speakers of this anumeric language struggled mightily, and their errors increased in accordance with the number of nuts initially placed in the can. In other words, Gordon's results suggested that the Pirahã were capable of accurately and repetitively matching quantities in their heads, but only if the number of items being mentally tabulated did not exceed three. For quantities greater than three, errors predictably worked their way into the responses. The Pirahã responses reflected a reliance on approximation or analogue estimation, rather than discrete quantity differentiation.

The erroneous quantity selections made by the Pirahã were not random. Instead, what Gordon observed was a clear correlation between the number of stimuli presented and the range of the typical Pirahã errors. For tested quantities of greater magnitude, the average deviation from the correct responses increased proportionally: the larger the target number was, the larger the errors became. This correlation was extremely consistent.[7] One might wonder why the observed pattern in the Pirahã incorrect responses is so pertinent. It is for two reasons. First, the observed correlation suggests that the Pirahã were in fact trying to mentally match the observed quantities. They were not simply giving up when quantities exceeded three. Nor were they randomly guessing because they were bored, for instance. Second, while the correlation suggests they were mapping quantities in their heads with those presented to them, they were obviously doing so in a fuzzy manner. This is exactly the manner we would predict given what we know about the number senses humans are innately equipped with.

A clear picture was painted by these initial results gathered among the Pirahã. These people struggle with simply differentiating quantities greater than three. They can do so, but in a fashion that relies on estimation. Similarly, the people clearly are capable of exactly differentiating small quantities. What the Pirahã appear to lack is the means of unifying these two genetically endowed capacities, even though in all respects these people seem normal genetically. They have been successful in their local ecology, adapting and surviving along the Maici River for, at least, centuries. There is no easy explanation for their demonstrable difficulties with basic quantity-recognition tasks, besides the fact that they speak a completely anumeric language and have no other numeric practices in their culture.

Gordon's results were widely discussed and were taken by many as a clear indicator, perhaps the clearest to that point, that some seemingly basic mathematical concepts are not wired into the human condition. They are learned, acquired through cultural and linguistic transmission. And if they are learned rather than inherited genetically, then it follows that they are not a component of the human mental hardware but are very much a part of our mental software—the feature of an app we ourselves have developed.

Since replicability is a fundamental tenet of any scientific enterprise, various cognitive scientists were eager to follow up on Gordon's work among the Pirahã. Just a few years after the publication of his seminal article, a team of cognitive scientists, including Michael Frank (of Stanford University) and Ted Gibson (of MIT), did just that. As described above, this team demonstrated that Pirahã terms for quantities have inexact reference. Their work was not limited, however, to the experimental corroboration of the claim that Pirahã is truly anumeric. It also included the replication of several tasks conducted by Gordon, more specifically the line matching, orthogonal matching, brief-presentation, and nuts-in-a-can tasks. The tasks were replicated in a village known as Xaagiopai (pronounced "ah-gee-oh-pie"), which is located several days downriver (by canoe) from the villages in which Gordon conducted his research. The researchers utilized different stimuli for their matching experiments, since they were concerned that the AA batteries utilized by Gordon might have rolled occasionally, making an already difficult task more challenging for the Pirahã. Instead, they used other uniform nonnative objects that the people are familiar with: spools of thread and empty rubber balloons. The spools of thread could be placed vertically on a table, without rolling. Fourteen Pirahã adults were presented, individually, with quantities of spools of thread on a table immediately in front of them and were then asked to match the number of spools with an identical number of empty rubber balloons. Excepting the different stimuli used in this study, the quantity-matching experiments were modeled carefully after those in Gordon. For the orthogonal, brief-presentation, and nuts-in-a-can matching tasks, the Pirahãs' performance was just like that observed by Gordon. So the researchers concluded that the Pirahã are incapable of exactly differentiating quantities greater than three when they have to spatially transpose the items or when they have to recall the number of items after seeing them briefly. Notably, these tasks are fairly trivial for populations familiar with number words and counting, such as adult English speakers.

In the case of the basic line matching task, without recall or physical rotation required, Frank's team did not in fact replicate Gordon's results. Instead, they found that, with some exceptions, the Pirahã were capable of precisely reproducing the quantity of viewed spools of thread, if those spools were simply presented in a line and if the participants were permitted to view that line throughout the task. These results presented a complication to the previous interpretation of Pirahã numerical cognition. The research team concluded that, while number words serve as a "cognitive technology" that is pivotal for the manipulation and recall of quantities, it is not a technology required for the mere recognition of quantities. The researchers suggested a possible alternative reason that the Pirahã struggled with the simplest one-to-one line matching task in Gordon's study: maybe the AA batteries had in fact rolled occasionally, and maybe this rolling complicated the people's quantitative perception.

During the summer of 2009 on another humid Amazonian afternoon, I found myself reading this more recent work on the Pirahã while conducting research with another indigenous group. Although convinced by the results of Frank and colleagues, I was also unconvinced that any of Gordon's results were due to rolling batteries. Gordon's work had, after all, been conducted with assistance from my parents (like the latter study), who served as translators and facilitators for his field research in the early 1990s. As someone who occasionally followed my parents to the jungle during my early teenage years (though we then lived in the United States), I had observed Gordon conduct some of his experimental work. And, much as I had observed the Pirahãs' difficulties with basic quantity recognition during those lamp-side sessions in the early 1980s, I saw them struggle with the similar tasks used in Gordon's work, regardless of the type of stimuli used. And maybe more crucially, I was aware that the Xaagiopai village used in the follow-up study was unlike the other Pirahã villages in an important respect. In the months prior to the experimental work undertaken there, my mother, Keren Madora, had conducted math lessons with the people. In contrast to the previous futile attempts at teaching basic numerical concepts, some progress had apparently been made when she resorted to innovating made-up Pirahã words such as *xohoisogio,* meaning 'all the sons of the hand.' Some of the people in that village seemed to have learned some basic quantity-recognition skills, at least in part due to such lessons.

To better resolve this issue, I returned to a separate Pirahã village several weeks later, and again in the summer of 2010. The fruit of that research, conducted with my mother—who speaks Pirahã extremely well, like my father—was a series of findings that corroborated most of the results obtained in both previous experimental studies of the people. The findings also offered unequivocal support for Gordon's strong initial claim that, absent number words or other symbols for quantities, untrained Pirahã do not consistently differentiate quantities greater than three. This is true even when the people are only asked to perform the most basic one-to-one line matching task.[8]

Our study replicated the aforementioned matching tasks, using the identical methods and stimuli as those used by Frank and colleagues, namely, spools of thread and rubber balloons. We also used other stimuli, with the same results, in follow-up work. However, since we tested Pirahã in a new village far from Xaagiopai, our participants had not been exposed to the teaching of number words in the months leading up to our experimental work. Fourteen adult subjects (eight women, six men) participated in our project, though we also conducted some tasks with children, who were eager to get involved. For all three matching tasks, the responses of the Pirahã contained errors when they viewed 4 or more stimuli. The ratio of correct responses fell from 100 percent for 1, 2, or 3 presented items, to about 50 percent for 5 items, and progressively less for larger quantities. For 10, the highest quantity tested, the responses were correct only about one-quarter of the time for the one-to-one line matching and orthogonal tasks, and about one-tenth of the time for the hidden matching task.

Simply put, the Pirahã struggle with the exact differentiation and recall of quantities in experimental contexts, whenever the quantities being tested exceed three. All three experimental studies support this conclusion, and the results in all three studies are remarkably consistent for most of the tasks. Furthermore, the relevant discrepancy in the results of Frank and colleagues is readily explainable. And, even if one is skeptical regarding the proposed explanation for that discrepancy (i.e., that it is due to number-training at Xaagiopai), the fact remains that all three sets of findings demonstrate that the Pirahã face difficulties with quantity-recognition tasks that are known to be simply solved by numerate peoples. The most plausible interpretation of the Pirahã data is that, without learning number words and associated counting strategies, the people lack the ability to fully unite the two innate capacities we all share for discriminating quantities. The two number senses remain dissociated, and quantitative cognition is limited when compared to that of numerate populations. Apparently the unification of these capacities requires that one be raised in a numerate culture and have practice with number words.[9]

In addition to replicating previous studies among the Pirahã, we subsequently tested their quantity-recognition skills in other ways. For instance, for some tasks they were asked to repeat a series of gestures, or a series of noises like clapping. In all cases, their performance still indicated difficulty with the precise differentiation of quantities. They apparently use their approximate number sense when discriminating quantities, regardless of the way in which the relevant stimuli are perceived.

Unfortunately, the results gathered among the Pirahã are still frequently misinterpreted. As noted previously, one facile and clearly incorrect interpretation is that the people's performance is somehow indicative of population-level cognitive deficiencies. This interpretation is not tenable from an empirically grounded perspective. Another easily disprovable interpretation is that the people simply make no effort to perform the relevant experimental tasks, or are paying attention to other things during the tasks. This interpretation is irreconcilable with a careful inspection of the results, since the subjects do not simply guess but, instead, consistently approximate correct responses. In other words, they are clearly paying attention to quantity, but in a fuzzy way. A third, more reasonable, interpretation is supported by all the data so far obtained among the people: they are anumeric, lacking numeric language or any other trace of numerate culture. It is this anumeracy that yields a clear impact on their discrimination and recall of quantities. This interpretation is supported not just by the experimental work but also by the reports of many outsiders who have experience interacting with these people. Nothing about indigenous Pirahã culture, whether behavioral or material, suggests the existence of precise enumeration. Their housing structures, their hunting implements, and their other smaller artifacts do not require precise quantity differentiation in their production.

Surely, though, some other facet of their culture requires the consistent discrimination of larger quantities? In that vein, some misunderstandings that I have come across at various times, in lectures and in other contexts, are evident in the following paragraph offered by a commenter online, in response to an article in *Slate* about the results of this experimental work: "If a woman of this people has seven, or even five, children, how does she raise and care for them, if she can't use arithmetic to know their relative ages? Can a mother even remember that she has more than two or three children at all? If she can, then she should be able to count in other contexts too."

Such comments reflect two key misconceptions about the human experience in a numerical void. First, tracking age approximately does not require the usage of number concepts. I can be confident that one relative is older than another simply because the first relative was alive when the other was born. And if there are three or more relatives, I can keep track of who is oldest by some simple syllogisms, or by direct comparisons between any two of the relatives.

So, if a mother has four children, she will know who was alive before all the others, and who was born when all the others were already alive. No indication of quantity is required to understand such concepts. The understanding of absolute ages, such as the number of trips a person has made around the sun (see Chapter 1), does require quantity differentiation. Yet this is a distinct concept from relative ages, though it seems tricky for many people to grasp this distinction—likely because they are embedded in cultures in which age and numbers are inextricably linked from an early age.[10]

The second point evident in the reader's comments is, superficially at least, more insightful. How can a mother not recall how many children she has? Surely all mothers must remember all of their kids at all times? Of course, but this is actually irrelevant to quantity recognition and counting. Say you come from a large family and go home during the holidays to visit your four siblings, Cory, Angela, Jessica, and Matt. Cory's flight is cancelled, however, and he does not make it in time for a planned family dinner. Had you not been forewarned of the cancellation before the dinner, when you sat down to eat you would immediately wonder, "where's Cory?" You would not, I suspect, look around and exclaim, "I see three siblings here, but I know I have four!" In other words, you do not need any notion of exact quantity recognition to appreciate that a loved one is missing. We perceive family members as individuals, not as faceless countable objects. Of course, we *can* count our family members. But we do not *need* to in order to recognize their presence or absence. And neither do the Pirahã. There is no evidence that suggests these people require precise quantity differentiation to remember who is missing, or for any other task in their culture. Were that the case, they would no doubt have adopted number words to facilitate such differentiation. A Pirahã child is remembered as an individual, not a number.

Questions regarding people without numbers often reveal something about the numerate ones doing the asking. We are so enmeshed in the world of numbers that it is hard for us to conceive of life without them. Our cognitive and material lives are not detachable from the numbers we have developed and which we impose on ourselves so early in life. The results obtained among the Pirahã could be, and sometimes are, used to exoticize this group of indigenes, to treat them as some sort of contemporary Paleolithic group. Such exoticization misses the true revelation offered up by such research. And the revelation is not about the Pirahã, but about all of us: humans need numeric language and culture to exactly differentiate and recall most specific quantities. Rather than being the simple by-product of innate mechanisms or naturally occurring processes, some seemingly basic quantity-recognition skills are acquired only through culture and language.

Thankfully, this conclusion is not dependent entirely or even principally on the results obtained among the Pirahã. Work on the development of numerical cognition among children in large industrialized societies has arrived at the same conclusion, as we will see in Chapter 6. Additionally, so has research carried out among another fascinating group of Amazonians—the aforementioned Munduruku. The Munduruku are a much larger indigenous culture. In contrast to the approximately 700 Pirahã, there are more than 11,000 Munduruku. They reside about 600 kilometers from the easternmost portion of the Pirahã reservation, on their own large reservation. They do share some commonalities with the Pirahã, despite different lifestyles. Historically, they share a fearsome reputation for having defended their native territories upon the arrival of Europeans. This reputation was also earned by another large group known as the Mura, of which the Pirahã were a small affiliate. As noted previously, the Munduruku also share a long-held reputation for struggling with numerical concepts.

Unlike the Pirahã, however, the Munduruku are not completely anumeric, since they do have number words used to denote sets of 1, 2, 3, or 4 items. However, close inspection reveals that these numbers are not as precise as the number words in most languages, falling somewhere in between English number words like *one* and *two* and Pirahã words like *hói* and *hoí*. This imprecision has also been corroborated experimentally. When asked to provide a number term for the set of dots randomly presented on a laptop computer screen, Munduruku speakers offered their word *pug* in nearly 100 percent of cases when one dot was presented, and used the term *xep xep* in 100 percent of cases when two dots were presented. Clearly, their language has words for 'one' and 'two.' It also has words that are typically used to denote sets of three and four items, respectively, but these terms do not refer to three and four items in all cases. This suggests that they are imprecise number words of a similar ilk as those used by the Pirahã. For quantities greater than four, other approximate words are used. These are best translated as 'some' and 'many.' Furthermore, the number words that exist in the language are apparently not used often, so numeric reference is uncommon. (In contrast, some languages in Australia and elsewhere also have modest sets of number words but make frequent reference to concepts like singular, dual, and plural.)[11]

Pierre Pica, a French linguist who conducts experimental work in Amazonia, has explored basic numerical cognition among the Munduruku along with a team of well-established psychologists. Together, the team has demonstrated that, like the Pirahã, speakers of this language rely on their approximate number sense to perform basic mathematical tasks when those tasks involve quantities exceeding three. In one study, Pica and company conducted four basic mathematical tasks with fifty-five Munduruku adults and ten French-speaking control subjects. Two of these tasks were approximate in nature, for instance, speakers were asked to quickly decide which of two dot-clusters on a computer screen contained the most dots. The performance of the Munduruku and French speakers was similar for these tasks, since they only required approximation rather than exact quantity discrimination. For two other tasks, however, the participants were asked to demonstrate precise quantity representation. Both tasks entailed the subtraction of dots. Participants watched as dots were "placed" into a can depicted on the computer screen, and continued watching as a number of the dots were "removed" from the can. For one task they were then asked to name the quantity remaining in the can, and for another they were asked to choose, among a set of images, the can with the correct number of remaining dots inside it. So, for instance, if they viewed five dots enter the can and four dots leave the can, they then viewed three "choice" cans simultaneously—a can with zero dots in it, a can with one dot, and a can with two dots. In this case the correct answer is, obviously, the can with one dot.

In the tasks requiring the exact recognition of quantities, the Munduruku responses were very Pirahã-like. There was a clear disparity between the results obtained with them and those obtained with the control group of French speakers. For Munduruku children and adults, performance was nearly perfect when the initial number of dots placed into the can was less than four. When the number of initial dots was four or greater, however, performance fell dramatically. The "Munduruku still deployed approximate representations . . . in a task that French controls easily resolved by exact calculation."[12]

While other research has explored numerical cognition among members of unrelated indigenous groups with limited number systems, the results obtained among the Pirahã and the Munduruku are particularly relevant to our understanding of the transformative capacity of numbers as conceptual tools, since these groups generally lack precise number words *and* grammatical number. (Recent research suggests that grammatical number is also helpful in acquiring number concepts.)[13] Unsurprisingly, the people also lack counting routines that are apparently crucial to the development of precise numerical concepts in the minds of children in numerate societies. (See Chapter 6.) The findings from these two unrelated Amazonian groups are remarkably similar, pointing to a clear implication: humans in cultures without any precise systems of numeracy struggle with exactly differentiating quantities larger than three. This conclusion helps inform our understanding of the effects that numbers have had on us as a species. After all, these results do not just reflect something about these two groups, but more tellingly, they reflect how basic mathematical cognition works in the absence of the conceptual support afforded by number words and associated symbols. Healthy adults without such symbolic tools rely on approximate estimation when handling quantities greater than three. Far from being hardwired somehow, exact representations of quantities such as 5, 6, and 7 are generally acquired through culture. They are a linguistically dependent phenomenon. Only when we are embedded in a matrix of culturally inherited symbols, such as number words, can we exactly and consistently distinguish such quantities.

In his book *The Number Sense,* famous psychologist Stanislas Dehaene (a member of Pica's team) offers the following interpretation of the quantity-recognition findings gathered among the Munduruku: "Our experiments . . . argue forcefully for the universality of the number sense and its presence in any human culture, however isolated and educationally deprived. What they show is that arithmetic is a ladder: We all start out on the same rung, but we do not all climb to the same level."[14]

Dehaene is of course correct that the Munduruku and Pirahã results support the universality of our innate number sense, or more precisely our two number senses. After all, both populations can consistently distinguish 1, 2, and 3, due to their exact number sense (technically the innate parallel individuation system). And both groups estimate larger quantities, using their approximate number sense. What is probably more remarkable about the relevant studies, though, is that they suggest that climbing *any* rungs of the arithmetic ladder requires numbers. How high we climb the ladder is not the result of our own inherent intelligence, but a result of the language we speak and of the culture we are born into. If somebody speaks an anumeric language, she or he has little chance of climbing, and probably little reason to climb. Numeric language and associated numeric practices enable seemingly basic kinds of quantitative thought.

Much research remains to be done before we fully understand the role that numeric language and other facets of numerate culture have played in shaping the mathematical cognition of the vast majority of humans alive today. This exploration will elucidate not just the role of numbers in the unification of our two basic number senses but will also illuminate other facets of quantitative thought. Work among the Munduruku has begun exploring, for instance, how numbers influence people's capacity to halve quantities. That research suggests the Munduruku are capable of halving quantities approximately, even when they have not previously been exposed to those quantities.[15]

Another issue recently investigated among the Munduruku is the extent to which these people use spatial concepts when thinking quantitatively. As noted in Chapter 1, people often use the physical domain of space to make sense of more abstract cognitive arenas. I noted in that chapter that nearly all cultures use space to make sense of time. Similarly, humans often construe quantities in spatial terms. This transfer of quantities onto space is utilized in our school systems, of course, in which students are taught to use number lines and the Cartesian plane. But a less systematic use of space seems to surface in many other cultures. Furthermore, some evidence suggests that young children and infants learn to map numbers onto space well before any schooling takes place.[16]

To better understand the cognitive mapping of quantities onto space, Dehaene and colleagues conducted the following task with the Munduruku. Participants were presented with two circles separated by a horizontal line. In the left circle was one black dot, and in the right circle were ten black dots. The participants were then provided a separate set of dots, for instance, a group of six dots, and asked to locate this quantity of dots along the horizontal line. Now, if the Munduruku do not perceive quantities in terms of space, their responses might be random. By way of contrast, Americans generally place quantities at regular intervals along the horizontal "number" line. For instance, they locate nine dots close to the circle with ten dots, and place two dots adjacent to the circle with one dot. The Munduruku also placed the dots in a regular fashion, but not in such a way that quantities were so neatly separated along the number line. Instead of such a straightforward linear approach, which seems to be learned in school, the Munduruku generally adopted what is called a logarithmic strategy for the task. Smaller quantities were more displaced along the horizontal dimension compared to larger quantities. Three dots were placed relatively close to the midpoint of the line, and nine dots were typically placed about twice the distance (compared to three dots) from the left circle. In the linear configuration most of us are familiar with, we would expect nine dots to be placed three times as far away from the left circle, when compared to three dots. In a logarithmic configuration, however, nine dots should be placed two times as far away as three dots, since $3^2 = 9$. Since the Munduruku are largely anumeric, their dot-placement strategy suggested to some that all people, regardless of their culture or language, use space to make sense of quantities. Anumeric adults without schooling do apparently use a mental "number" line, though their mental line is structured logarithmically.[17]

But now it appears this may not be the case in all unschooled populations. Recent results obtained with an indigenous group on the other side of the world suggest that a human number line may not be universal after all. Led by Rafael Núñez of the University of California, San Diego, a team of cognitive scientists replicated the experiments on mental number lines with the Yupno, who live in a remote mountain range of Papua New Guinea. Although the Yupno have number words, these people do not measure space or time in precise ways. And, it turns out, they do not mentally map quantities onto space in a predictable manner. Unlike most Munduruku, they show no predisposition to place quantities along some mental number line, logarithmic or not. Instead, when asked to place given quantities of dots and other stimuli along a line, they consistently chose the endpoints of the line. This failure to divide the line in accordance with quantities suggests that not all humans think of quantities in spatial terms. Instead, some people perceive quantities in different ways that are at least partially contingent on the conceptual rigging they acquire from their native culture.[18]

Obviously, we still have much to learn about how numeric language and other parts of numeric culture shape humans' conception of quantities. It remains uncertain, for instance, how much cultures differ in their propensity for using mental number lines. One thing is clear from our discussion in this chapter, though: members of diverse indigenous cultures, speaking unrelated languages in remote jungles, are helping reshape our understanding of mathematical thought. Studies with these populations have now shown that numbers and counting are essential for even seemingly modest quantitative cognition. The studies have demonstrated that, even though all people are born with certain very basic arithmetic abilities, only members of cultures with numbers can truly start climbing the arithmetic ladder.

No Sounds and No Numbers

The Munduruku and the Pirahã are not the only people with anumeric, or primarily anumeric, communication. There is another group of people that lacks numeric language and is also shedding light on the way in which numbers impact basic mathematical thought. The people in question are homesigners in Nicaragua. These are deaf Nicaraguans who have, for a variety of reasons, never had the opportunity to learn a sign language. Instead, like homesigners in other parts of the world, they use innovated hand gestures to speak with those around them, particularly family members who co-learn and co-create the innovated homesign systems. Homesigners are a testament to the power of the human need for communication—in cases in which children lack hearing and exposure to a fully developed language, elaborate communication still arises. Yet this elaborate communication, at least in the case of Nicaraguan homesigners, lacks number words.

In a fascinating study, psychologists conducted a series of experiments on the numerical cognition abilities of four adult Nicaraguan homesigners. These adults lack any knowledge of number words. Unlike the Pirahã and Munduruku, however, they are embedded in a numerate culture. They have knowledge of the importance of differences between quantities. For instance, the homesigners can recognize the approximate values of monetary notes and are capable of differentiating smaller and larger bills. Yet, despite such an awareness, and despite their awareness of the existence of precise quantities beyond three, they do not have their own means of referring to such quantities in precise ways. They lack numbers.

Researchers tested the extent to which this absence of numbers impacts these Nicaraguans, who, like the Pirahã and Munduruku, show no signs of congenital cognitive defects. The study sheds further light on the mathematical abilities of healthy anumeric adults, but does so with a group of people that have been made aware, throughout their lives, of the existence of numbers. Those numbers have simply not infiltrated their cognitive lives via the acquisition of number words and counting routines. While the homesigners do use gestures to communicate about quantities, they do so in an imprecise manner when larger quantities are involved.

After running a battery of experiments, including a basic one-to-one line matching task similar to that conducted several times by now among the Pirahã, the researchers arrived at a strikingly similar result. The homesigners "cannot reliably make the number of items in a second set match the number in a target set if the sets contain more than three items."[19] Like the Pirahã and the Munduruku, these adults struggle with exactly differentiating and replicating sets of items when they exceed three. For instance, the homesigners viewed cards with specific quantities of items depicted on them. They were asked to demonstrate, with their fingers, how many items were on the card. When only one, two, or three items were depicted on the card, they presented the correct number of fingers in all trials. For larger quantities, their proportion of correct responses plummeted, and the magnitude of their errors increased in accordance with the quantity of items. In contrast, deaf Nicaraguans who speak a formal sign language and who know number words were unerring, as were nondeaf Nicaraguans who know Spanish number words. In short, the results with these Nicaraguans converged on a by-now familiar conclusion: humans need practice with number words to consistently and exactly differentiate quantities greater than three.

Conclusion

If numbers are so useful cognitively, why have some groups chosen to abandon them or failed to adopt them? We could offer deliberately anodyne responses to this question, for instance saying that "these people are better off without numbers," but such responses run a real risk of paternalism or misplaced cultural relativism. The true answer to this question is likely lost in the unwritten histories of the cultures in question, at least in the case of the Munduruku and Pirahã. Surely such groups could benefit from the advantages afforded by the adoption of these wondrous cognitive tools. Yet, just as surely, these groups have long survived and excelled in their ecologies without the assistance of numbers.[20]

In light of the usefulness of numbers and their near ubiquity in the world's languages, it is in one sense startling to find that anumeric populations exist. Yet, inevitably, the global inventory of languages offers exceptions to our expectations that certain linguistic features should be universal. Cultures and languages differ radically. Given this fact, it is in another sense less surprising that some people live in worlds without number words, numerals, conventionalized numeric gestures, or other signs for precise quantities. I have tacitly implied in this chapter that, in the context of our present quest to understand how numbers transformed the human experience, this is a useful thing. After all, groups of healthy anumeric adults offer us an invaluable window into the nature of human quantitative thought. They offer clear evidence that, without numbers, we cannot build on our innate capabilities in order to make perfect sense of all quantities.

6

QUANTITIES IN THE MINDS OF
YOUNG CHILDREN

While we trace our age and existence to the moment we were expelled from our mothers' wombs, this is largely a matter of convenience. In reality we are weaned off of unconsciousness, brought to life gradually in utero. This weaning is limited by the nature of our confines, which restrict most stimuli from touching on our mental lives. But the womb cannot hold back all stimuli, and it is in there that we begin to contact mind-external physical stimuli that profoundly impact our cognitive and behavioral lives henceforth—our fingers. These stimuli are experientially basic, seeping their way into our sensory experience, even our mouths, before we are aware of smells, sights, and sounds (with few exceptions). The first time I saw my son's face, it was depicted in a three-dimensional ultrasound rendering. Some two months before his birth, his fingers floated in the space right next to his cheeks, seemingly a reassuring constant in his newfound existence. Before he saw true light, he was "seeing" the fingers around him, touching these individuated portions of his own body, apparently feeling them one at a time. The same is true of other babies in the womb, of course. This prebirth phase of life is just the beginning of a long cognitive journey taken hand in hand with our fingers. The quantity of fingers on our hands plays a foundational role in our distinctly human numerical thought, evident in how the world's people have so frequently named numbers after hands. Yet, before we can consider that foundational role, we need a clearer understanding of those numerical concepts that may predate our awareness of our countable fingers. Because, though we encounter our fingers so early on in life, we are apparently aware of some numerical distinctions even prior to such an encounter. After all, humans are prewired with the numerical neurocognitive systems mentioned in Chapters 4 and 5: the approximate number sense and what I have termed the exact number sense. Together these number senses allow us to perceive quantitative distinctions very early in life.[1] Even with these senses at their disposal, though, it is difficult for children to learn how to think precisely about all quantities. Learning how to make numerical distinctions is not easy and depends largely on the acquisition of number words. For most people anyhow, those words, so often finger-based, serve as the gateway to the awareness of integers.

How Babies Think about Quantities

As we have seen, the exact number sense is clearly evident among anumeric populations without any means of linguistically transferring numerical concepts from each generation to the next, or within generations. Similarly, all adult human populations, even anumeric ones, possess the ability to discriminate larger sets of items numerically if the difference between sets is great enough. All neurocognitively normal adults can differentiate 6 items from 12 items, or 8 items from 16 items. This ability is evidence of our approximate number sense. Number words and counting strategies allow us to exactly differentiate all quantities, including larger ones, essentially by fusing together our two innate number senses.

But how, one might reasonably ask, can we tell that the two senses in question are actually innate abilities? Other features of the human cognitive toolkit that were once presumed to be innate now do not appear to be. For instance, many linguists have concluded in the past that the origin of language itself was due primarily to a mutation, or series of mutations, in the human genome, and that all humans share a 'language instinct.' In other words, humans with a specific genetic mutation (or series of mutations) that enabled language were selected for by nature, since that genetic characteristic was so beneficial to reproduction. Fewer linguists maintain that position today, and language is best considered by many scholars to be a collage of culturally variable but often similar strategies for communication and information management. From this increasingly popular perspective, natural selection likely resulted in a suite of cognitive and social abilities that helped to foster language, but did not directly produce a specifically linguistic instinct. How can we be so sure, then, that our number "senses" represent specific quantitative instincts and are not also the result of common convergent strategies for thinking about quantities? How can we be confident that we actually exhibit innate numerical thought?[2]

It is difficult to offer a definitive answer to this question. There are two major components to the answer, however; two general kinds of evidence that have left us with a relatively strong certainty that humans have a number instinct or, more accurately, two innate capacities that we can use to discriminate quantities. One kind of evidence is gathered from the behavior of other animals. As we will see in Chapter 7, many species share similar capacities for the exact differentiation of small quantities as well as the approximate discrimination of higher quantities. So we can state with confidence that the two numerical systems in our brains are phylogenetically primitive. This means they have been around for millions of years in human brains and in the brains of our precursor species. They can be traced back to other extinct species whose descendants include humans and many related vertebrates. Another source of evidence is the behavior of young, prelinguistic children. This latter sort of evidence suggests that some mathematical abilities are ontogenetically primitive. In other words, they are abilities we bring to the table before we start experientially acquiring concepts during childhood. They are a genetic gift. Just because the two number senses in question are genetically gifted to us does not imply the absence of cross-cultural variation in how these capacities flourish in our minds as we age, or in how we manage to exploit these systems for fuller mathematical thought. Much work remains to wholly understand the nature of quantitative thought in human infancy across the world's cultures. Yet much work has already been done, and here I draw attention to a few of the seminal findings on this topic.

But first, a simple question you might have: How can we figure out what is going on in the minds of infants, when they are prelinguistic and cannot be told what to do in experimental contexts? This is a tricky issue methodologically and one that took some innovation to overcome. Overcoming the obstacle resulted in a quasi-revolution in our understanding of infant numerical cognition in the past 30 or so years, simultaneously calling into question previous results that may have been influenced by unfair expectations of infants' intuitions regarding the goals associated with particular experiments. The obstacle was overcome as researchers began relying on tasks that required little in terms of infant physical participation and interaction with experimenters, and focused more narrowly on what infants pay attention to in experimental contexts. This focus on attention makes sense given that humans, like other species, fixate on novel stimuli.

Consider a quotidian environment you are likely very familiar with: a crowded restaurant. When you enter the restaurant, you may notice the din of the conversations, the clink of silverware contacting plates, glasses being set on tables, and so forth. This is what you expect from such an environment, and as you sit and eat, such non-novel stimuli will fail to maintain your interest. You and the remainder of the restaurant's patrons will continue to eat and drink, focused on your meals, your own conversations (hopefully these have novel stimuli mixed in, or they are likely to dull your interest as well). Now consider what happens if a novel stimulus enters your perceptual sphere, for instance a glass rolls off a server's tray and lands on the floor, shattering. Your attention is immediately drawn to the sound of shattering, as you seek to discriminate the source of the noise you just heard. When your attention is altered in that manner, several things happen physically. Your gaze is drawn to the noise, and likely all heads in the restaurant will momentarily swivel to focus on the novel stimulus. Less visibly, the patrons will probably stop eating for a moment, suspending activities like swallowing. Crucially, these propensities for gaze fixation and ingestion pausing are developmentally basic, traceable back to our infancy. As a result, child-development researchers have for some time realized that they can tell when babies consider stimuli to be novel or not. So, when examining the attention of infants, the researchers can generally observe whether those infants recognize a new stimulus—whether that stimulus is a new color, a new shape, or a new quantity. To make such an observation, researchers simply have to track whether there are any associated changes in the babies' gaze or their ingestion patterns while they are being presented with stimuli. In practice, the experimenters examine patterns in babies' staring and sucking during particular tasks. Staring and sucking are not straightforward to measure precisely, and methodologies based on the measurement of these behaviors required the advent of new tools made available in the past few decades. These included electronically monitored pacifiers and video capable of tracking the gaze and eye movements of children.

Now let us consider some important experiments on infant numerical cognition, all of which are based on the assumption that infants stare longer at particular kinds of stimuli. We should begin with a now-famous study conducted by psychologist Karen Wynn. The results of this study were published in *Nature* more than 20 years ago, and the experiments in it have been refined and replicated in various ways in the intervening years. Wynn's influential study is a logical starting point for us, since it strongly suggested that infants recognize discrepancies between 1, 2, and 3, even in their preverbal stage. The infants in question were 5 months old on average. More recent studies have examined the numerical cognition of younger infants, including newborns. (We will consider one such study shortly.) Wynn recruited thirty-two infants to participate in the study. Half the infants were assigned to participate in a task that tested their ability to add $1 + 1$, while the other half participated in a task that tested their ability to subtract $2 - 1$.[3]

So how did Wynn test these 5-month-olds for the two tasks? The following ingeniously simple methods were used: Infants were placed individually in front of a display case, which contained a doll-like figure that naturally attracted their attention. Additionally, the display case had an opaque screen that could be lifted up to block an infant's view of the doll-like object on display. After this screen was raised, the figure was no longer visible. Crucially, though, there was a gap next to each side of the screen in the display case. After the infants' attention was first drawn to the figure, the screen was raised to block their view of it. Next, the researcher conducting the experiment placed their hand through a side door in the display case while holding another doll-like figure, identical to the one hidden from view by the screen. The infants could see the hand, along with the item it held, through the gap between the screen and the side of the display case. From the infants' perspective, now a second item was being added to the other identical doll-like object behind the screen. Here is where a key methodological trick came in: the display case was equipped with a secret door through which the researcher could remove the original item, out of view of the infants, while the second identical item was being added. So, though the infants viewed one object being added to another object, in actuality the original object could be removed simultaneously from behind the screen, unbeknownst to the babies. Finally, the screen was lowered after the second object was added. Under a "possible outcome" condition, this lowering revealed two identical toy-like objects, in keeping with what the infants had perceived. Under the "impossible outcome" condition, the screen lowering revealed only one doll-like object, since the original object had been trickily removed through the display case's secret door.

Given that humans, including infants, stare longer at unexpected and new events, Wynn hypothesized that the infants in her study might stare longer under the impossible outcome condition. That is, if they saw one item being added to another identical item, but then discovered that the result of this addition was only one remaining item, they would be somewhat perplexed. In contrast, if two items were evident after the addition of the second, they would be unfazed by the expected result. Framed differently, if 5-month-old infants recognize that $1 + 1$ equals 2 and not 1, such an experiment should reveal that recognition. The straightforward prediction was that infants would stare longer at the display case in the impossible outcome condition, when $1 + 1$ appeared to equal 1 instead of 2. And they did stare longer. To a statistically significant degree, the infants gazed longer at the display case when the screen was lowered to reveal only one item.

For the subtraction task, the same display case and items were used, but the sequencing of events witnessed by the infants was essentially reversed. First, the infants' attention was drawn to two doll-like figures in the middle of the display case. Next, the screen was raised, preventing the infants from seeing either figure. Then an experimenter's empty hand reached into the display case, and the hand was again visible through the gap between the screen and the side of the case. The hand then removed one of the figures from behind the screen, in a manner clearly visible to the infants. Under an impossible outcome condition, the secret door was used to add an extra, identical figure while the visible removal took place. As a result, two figures remained after the screen was lowered, despite the fact that one figure had clearly been removed. Under a possible outcome condition, no additional figure was added through the trap door, and when the screen was lowered after the visible removal of one of the items, only one remained. For this subtraction task, the results were quite similar to those obtained for the addition task: infants stared longer at the seemingly impossible outcome. If they saw one of the two original figures being removed, they expected only one figure to remain. The infants appeared to recognize that $2 - 1 = 1$. In sum, the results of both tasks in Wynn's study suggest that prelinguistic infants can distinguish one from two objects. Since her path-finding study, newer methods have been utilized to explore this issue more carefully. As a result, it is now generally agreed that infants can in fact differentiate one through three items consistently.

The second infant cognition study we discuss appeared some 8 years after the publications of Wynn's study, and was coauthored by psychologists Fei Xu and Elizabeth Spelke. (Spelke, a researcher and professor at Harvard, is one of the more influential developmental psychologists in the world.) We address the study's findings here because it offers strong support for the existence of an approximate number sense, demonstrating lucidly that prelinguistic infants are capable of recognizing, in coarse ways, quantitative differences between large sets.[4]

Here is how the first experiment in Xu and Spelke's study worked. Sixteen infants, with an average age of 6 months, were habituated to a display of 8 black dots or 16 dots, on a white display. This means that infants were presented with stimuli until they were bored of them, or did not find them novel anymore. They were considered habituated when they stopped staring at the stimuli, or when they had seen fourteen consecutive arrays of dots. Under the 8-dot condition, alternating displays of 8 dots varying in size, configuration, and brightness were presented until the infants were habituated. Under the 16-dot condition, alternating displays of 16 dots, also varying according to parameters like size and configuration, were presented until the infants were habituated. For both conditions, infants were then presented with displays of 8 dots or 16 dots after becoming habituated to the initial quantity of dots. Under the first condition, the posthabituation arrays of 16 dots represented novel stimuli, because the infants had only been viewing arrays of 8 dots previously. Under the second condition, the reverse was true, and the posthabituation arrays of 8 dots were novel, since the infants had previously been viewing 16-dot arrays. Or at least we, as adults familiar with numbers, would recognize such arrays as being novel, because we know that 16 is not the same as 8. But what about infants who have never learned to count, nor to speak for that matter? Xu and Spelke's results offered compelling evidence that these infants, too, can recognize the difference between 8 and 16 items.

The results of the first experiment were straightforward: infants tended to stare a couple of seconds longer at displays that contained a quantity of dots distinct from the quantity to which they had become accustomed. So, if they had become used to seeing groups of 8 dots, they stared significantly longer at a group of 16 dots. Conversely, if they had become used to groups of 16 dots, they stared longer at a group of 8 dots. The visual attention of the babies demonstrated plainly that they recognized the difference between 8 and 16 dots, regardless of other variables, such as the size and configuration of the dots. In other words, even for large quantities most infants seem to be able to recognize numerical disparities, at least when the ratio between compared sets is pronounced.[5] This latter disclaimer is crucial, however, as evident from the results of Xu and Spelke's second experiment. For that experiment, the researchers replicated their first experiment with one crucial difference: They tested the infants' ability to recognize the disparity between arrays of 8 and 12 dots, instead of 8 and 16 dots. In this case, where the ratio of quantities was reduced to 2:3 (8:12) from 1:2 (8:16), the results shifted dramatically. The infants' staring patterns reflected no appreciation of the disparity between 8 and 12 dots.

These results, as well as those in other related studies conducted by developmental psychologists, offer compelling evidence that infants can recognize disparities between large sets of items, provided those disparities reflect at least a 1:2 ratio. This is evidence of an innate approximate number system, much as Wynn's work is evidence of a comparatively exact number system. Both systems serve as crucial cognitive precursors to the more refined numerical cognition of adults. As we saw in Chapter 5, though, such refined adult quantitative reasoning depends on linguistic intervention. Xu and Spelke noted as much when discussing our two innate numeric capacities. In their words, "as children learn the meanings of the number words and the purpose of the counting routine, they may come to bring together these two types of representation to form a unitary, distinctly human, and language-dependent notion of discrete number."[6] Which is not to suggest that this process of unification is simple or straightforward; in fact much debate persists among specialists as to how exactly the "bringing together" happens.

Such experimental research demonstrates that human infants can recognize some numerical disparities at a young age. The exact and approximate number senses hardly enable us to quickly and accurately solve most math problems, but they do give us a head start on such problems. Yet the studies in question do not demonstrate that human babies are equipped with truly abstract numerical concepts. Recognizing the difference between one and two doll-like figures, or between eight and sixteen dots on a screen, implies only that infants are drawn to visual numerical disparities. But, some might argue, such recognition does not imply that infants think of quantities in abstract ways—ways that are not contingent on visual perception. Phrased differently, infant quantity differentiation may not be cross-sensory or cross-modal. For instance, babies may recognize that two doll lions are distinct from one doll lion, in the visual modality. They may also recognize that two beeps played sequentially sound different than only one beep, in the auditory modality. But this physical appreciation of disparities in the two separate modalities does not necessarily imply that they recognize some connection between, say, two beeps and two lions. Such cross-modal recognition of quantity similarity would offer stronger evidence that what is happening in the minds of infants is numerical thought in a truer and more abstract manner. Unsurprisingly, perhaps, recent experiments have sought to test the cross-modal recognition of quantities by infants.

Some of these more recent experiments have also addressed another potential issue with prior work on the numerical cognition of infants: the advanced age of the subjects. This may seem an unusual issue to raise, given that the subjects in Wynn's study, for example, were only about 5 months old. Yet establishing the existence of a particular cognitive skill at such an age does not necessarily imply that it is instinctual. Such young-life evidence certainly supports claims of mathematical instincts, but it is unclear which quantitative skills infants begin to hone in culturally specific ways, even so early in life. And it is worth bearing in mind that the research of most developmental psychologists is dedicated to exploring the numerical cognition of infants in Western and industrialized cultures, so the results offer little information regarding any potential cross-cultural influences on mathematical thought in the first months of infancy.[7]

Yet the third seminal study we should examine, conducted by psychologist Veronique Izard and her colleagues (including Elizabeth Spelke), addresses both preceding points of contention. The captivating results of the study demonstrate that infants can recognize some disparities between quantities on an abstract, cross-modal, basis, and that they are capable of doing so shortly after birth. Izard and colleagues found parents willing to let their newborn infants participate in the study. In fact, they had to find many willing parents, since only a fraction of the infants selected for participation ultimately contributed to the study's results. Sixty-six infants were selected, but fifty of these infants were excluded from the actual sample due to fussiness or other problems, like falling asleep. This gives us some indication of how difficult it is to conduct such research! Those of us doing cross-cultural research in far-flung places like the jungles of Amazonia or the highlands of New Guinea may complain of the particular challenges we face when doing field research, but we are unlikely to encounter subjects that fall asleep in the middle of an experiment. The infants who ultimately participated in the study of Izard and colleagues were, on average, only 49 hours old. Such a young age allows us to rule out, with a high degree of confidence, the influence of their early-life experience. As we noted before, kids do encounter regular quantities, in the form of their fingers, in the womb. Furthermore, the mother's heartbeat and voice are audible to an infant prenatally, giving some familiarization with regular intervals of auditory stimuli. But it seems exceedingly unlikely that experiential and cultural factors impact the way numerical cognition develops in utero. Furthermore, Izard and colleagues' study demonstrated that humans are capable of visually discriminating quantities shortly after birth, and obviously they are not exposed to quantities of items visually in the womb.[8]

Izard and company offered clear evidence that newborn babies can use their approximate number system for the abstract comparison of quantities across modalities. Infants were played a series of syllables, such as 'tu-tu-tu-tu' or 'ra-ra-ra-ra.' These series contained a consistent number of syllables, each of which was followed by a brief pause. For instance, an infant could hear four syllables, followed by a pause, followed by four more syllables, and so on. Auditory stimuli were played for 2 minutes, during which time each infant became habituated to the quantity of syllables. After these 2 minutes, the infant was presented with images on a computer screen. These images contained arrays of a particular number of brightly colored shapes with mouths and eyes (to attract the babies' attention), which either matched or did not match in number the sequence of syllables the baby had just heard repetitively. Izard and her colleagues hypothesized that, if infants have an abstract cross-modal appreciation of quantities, then they should exhibit different responses to the visual arrays presented after the syllables. They should stare longer at certain arrays, depending on whether the arrays matched, in number, the series of syllables they had just heard. And that is exactly what the researchers observed. For instance, when played a sequence of four syllables, infants then generally stared longer at the screen when it contained four images than when it contained twelve images. They also stared longer at the screen when it contained four images than when it contained eight. (The staring discrepancy was greater in the first condition, though, since the difference between 4 and 12 is more pronounced than that between 4 and 8.) When presented with a sequence of six syllables, they subsequently stared longer when the screen had six images on it than when it had eighteen. The staring differences across matching and nonmatching conditions often exceeded 10 seconds—not a particularly subtle result. In short, the infants' staring patterns corresponded neatly to their relatively heightened interest in equal quantities, as though they were appreciating a newfound correspondence across modalities. Izard and company's findings further support the claim that humans are born with an ability to numerically approximate large sets of items, while also suggesting that this approximation ability is abstract and not simply tied to one particular sense like vision.

The three studies we have considered here demonstrate some of the innovative ways in which developmental psychologists are exploring the numerical thought of infants. The studies are congruent with the basic conclusion presaged prior to this discussion—humans have an instinctual abstract understanding of number, evident even shortly after birth. They have approximate and exact innate number senses.[9]

Kids and Counting

So infants seem to be pre-equipped with number senses at birth. Yet the existence of such senses only goes some of the way toward filling out our picture of how humans are capable of distinctly mathematical thought. In a way, the bulk of the mystery remains, since it is unclear how our innate senses are eventually co-opted for arithmetical thought. Given how basic those number senses are, how do we do the remarkable work of recognizing, for instance, that octopi have a specific number of tentacles, not just some fuzzy quantity? Such a basic quantitative discrimination is not afforded to us by our innate senses. So how do we get from point A, simple innate quantity discrimination, to point B, exact differentiation of all quantities? How do we truly step into the realm of natural numbers? One way to find the potential answer(s) to this question is to explore how children progress in their quantitative reasoning as they age. Many psychologists have conducted, and continue to conduct, such explorations. Here I survey a few illustrative findings uncovered during such explorations. The studies summarized here are, I think, representative of the kinds of methods being used to refine our understanding of how kids learn numerical concepts. But it should be borne in mind that there are literally thousands of published studies on this vast topic. The studies discussed here serve to demonstrate that the progression toward the awareness natural numbers is a very painstaking and gradual one. And it is a progression that requires repeated practice with linguistic stimuli.

Until the latter part of the twentieth century, the numerical skills of preschool children were generally underestimated. It was once believed that kids do not learn most basic numerical concepts until the age of 5 or so. Part of the evidence for this was the behavior of 4-year-old children on so-called conservation tests. In such tests, children are shown two lines of objects, say, a line of six glasses and a line of six bottles. The lines are presented in one-to-one correspondence, so that it is presumably clear that the two sets are equal in number. When asked which line of objects is greater in number, children generally reply that they are equal. Now, if the experimenter stretches out the distance between the objects in one line so that, for instance, the line of glasses is longer than the line of bottles, the children's responses may change. Their responses in early experiments on quantity conservation often indicated that they thought the number of items in the two lines differed after one line was merely elongated. They would state, for instance, that there were more glasses than bottles, though no glasses were added and no bottles were taken away. In other words, judging from such simple conservation tasks, kids under the age of 5 never recognized that quantity is maintained regardless of the length of the compared lines of stimuli. A change in the overall size of a set of stimuli appeared to confound kids' perceptions of the number of stimuli.

In actuality, though, young kids are at least sometimes capable of conserving quantities—they can recognize which of two arrays of items is greater in number, regardless of the length of the arrays. Some experiments have now demonstrated that the results obtained in early conservation tasks were due in part to confusion on the part of the children—confusion about the aims of the researchers. Put yourself in the place of a child in such an experiment for a second. If an adult, whom you presume understands much more about how things work than you do, presents you with two visually distinguishable arrays and asks you which one has "more," though they are clearly equal in number (given their one-to-one correspondence), how might you respond? It is hard to say, of course, but you might interpret their question so as best to make sense of their presumably pure motives, perhaps construing the line that has "more" to be the one that covers more space, rather than the one that has more objects in it. You might assume that the longer line must have more of something (like space) than the other, because otherwise the adult would be deliberately asking a misleading question, and why would an adult do that? In other words, such simple conservation tasks can potentially say little about the numerical cognition of children and may instead say more about how hard kids work to make social inferences during conversations with adults.

This latter possibility is supported by more recent research on this topic, which has now demonstrated that some young kids do recognize when one line has more items in it during conservation tasks. In fact, one study first demonstrated this point decades ago. The study introduced an interesting methodological innovation. Researchers presented lines of items the children wanted to eat, specifically, M&Ms. For example, preschool kids viewed two lines in one-to-one correspondence, each containing four M&Ms. Then the lines were altered, so that one of the lines was now shorter than the other while containing six M&Ms instead of four. Based on older findings, the children would have been expected to perceive the longer line as having more candies if they were asked to simply choose the line with more items. However, when the kids were instead asked to pick a row of candies to eat, to do a line of M&Ms (so to speak), the majority of kids chose the line with the greater number of candies, even though it was shorter. They would not have been able to do so if they did not recognize discrete quantity apart from spacing. Child psychologists have now found that young kids are more capable of figuring out this task than was once thought. Yet debate remains regarding the age at which kids are capable of such quantity conservation.[10]

To better understand how kids' numerical immersion takes place, consider the results of a more recent influential study on the development of children's cognition. (The study was conducted by Harvard psychologists Kirsten Condry and the aforementioned Elizabeth Spelke.) Through a series of experiments, the study demonstrated that when 3-year-old kids learn number words, they initially have only a poor grasp of the meaning of those words. In fact, this general finding has been supported by many recent studies: kids do not really understand the meaning of numbers when they first learn them. In the study in question, the researchers examined what number words actually represent to 3-year-olds, focusing on this age in part because 3-year-olds have been exposed to number words but have not had extensive math lessons. They can typically recite numbers from 'one' to 'ten.' The study's findings demonstrated that 3-year-old children have only a very basic understanding of the concepts associated with such number words. The kids understand, for example, that a label like 'eight' is used to describe a set of objects with a specific quantity. They also recognize that a word like 'eight' refers to a different set size than a word like 'two.'

Yet the study also revealed that 3-year-olds are not aware how many 'eight' is, nor whether it is always more than 'four' (among other failures). In other words, while 3-year-olds know how to recite number words from 'one' through 'ten,' they do not understand the associated concepts in the way that adults do. They have yet to recognize that these words refer to specific quantities. Eventually, kids' innate aptitude for discriminating 1 through 3 helps them make sense of the number words they hear all around them. They gain an appreciation of the true meaning of words like 'one,' 'two,' and 'three' at least in part because of their innate number sense. This appreciation then kickstarts the realization that other counting words also correspond to sequential, precise quantities. But the latter realization is only made gradually. It is certainly not the case that, when kids learn number words for most quantities, they are merely acquiring labels for concepts with which they are already familiar.[11]

Such research helps reveal how exact numbers are constructed in the minds of children over time, as they are exposed to number words and counting practices in their particular milieu. The consensus evident in developmental psychology is clear: quantity differentiation skills are innate in only coarse-grained ways, and they require linguistic and cultural scaffolding to be built up. As we saw in Chapter 5, such scaffolding is not universal in the world's cultures. But it is usually present. While populations vary with respect to the number bases they employ, or the extent to which they rely on formal mathematics, the vast majority of cultures have number systems and associated forms of counting. And these verbal counting strategies, often supported by finger counting or some other tally system, are crucial to the development of numerical concepts in the minds of children.[12]

While many of the specifics regarding childhood number acquisition remain cloudy, there are some well-established principles that children acquire after being consistently exposed to number terms and other technologies associated with counting. One pivotal principle learned by young children is the *successor principle,* typically acquired around the age of 4, which refers to the awareness that each number in a counting sequence refers to a quantity exactly one greater than the previously named quantity. An understanding of the successor principle implies that children have grasped that numbers do not just refer to amounts in arbitrary ways, but that number sequences are structured so that numbers label quantities that differ from their preceding number by one and only one.[13]

Another key waypoint on the route to arithmetical thought is referred to as the *cardinal principle.* When kids acquire this principle, they recognize that the last number said when counting items describes the cardinality or quantity of the entire set of items being counted. When children reach the stage of cardinal-principle knowing, they recognize that each number precisely describes a particular set size. This realization is gradual and not easily arrived at, and the time it takes to reach it varies from child to child. Still, kids go through predictable stages on their way to reaching it: first they are "one-knowers," recognizing that the word for "one" represents sets of one and only one item. Then they become "two-knowers," then "three-knowers," and only later cardinal-principle knowers. They become one-knowers, two-knowers, and three-knowers with comparative ease, at least in part because of their innate ability to keep track of 1–3 objects in a precise manner.

The cardinal principle helps children appreciate equinumerosity (i.e., that any two equally numbered sets of items can be placed into one-to-one correspondence with each other). This fantastic achievement requires months and even years, during which time children are bombarded with linguistic stimuli that help make the key realization possible. As we saw in Chapter 5, though, not all the world's children are so bombarded. Since children in some cultures are not exposed sufficiently to the necessary stimuli, they are unlikely to become familiar with the cardinal principle and to consistently recognize one-to-one correspondence between large sets that are identical in number. Framed differently, the results we observed with anumeric populations are predicted by the work of developmental psychologists who have illuminated the way in which kids in numerate cultures acquire numbers.[14]

The mechanisms through which number words and counting yield abstractions like the cardinal principle are still being fleshed out experimentally. Interestingly, recent work by a team of psychologists has shown that manual gestures help lead the way in the tying together of numerical concepts and words. The team included Susan Goldin-Meadow, a University of Chicago psychologist who has long been at the vanguard of the study of human gestures and associated cognitive processes. In a series of experimental tasks with 155 kids, the researchers examined the numerical gestures of children 3–5 years of age. They found that, for those children who still did not understand the cardinal principle, their usage of words to denote quantities was more limited than their usage of gestures to denote those same quantities. The children were better at referring to small quantities exactly with manual gestures rather than words. As the researchers note, their results demonstrate that "before children learn the cardinal meanings of the number words 'two' and 'three', they are able to access non-verbal representations of those set sizes and communicate about them using gesture."[15]

Furthermore, the psychologists found that children who do not yet know the cardinal principle are also better at approximately labeling larger sets of items with their fingers than with their words. Gestural representation of some numerical concepts appears to precede verbal representation. Perhaps this is not surprising, given the inherent advantages of the fingers in representing some quantities, when compared to words. Fingers can iconically represent the number of items in a small set via one-to-one correspondence, with separate fingers simultaneously used to represent each item. They can also more readily approximate larger quantities by the simple presentation of one and two hands, respectively. In contrast, words are generally arbitrary, must be memorized, and cannot be easily placed by kids in exact or approximate correspondence with a given set of objects.

This ontogenetic advantage of the fingers and hands is also evident in the crosslinguistic patterns in number terminology discussed in Chapter 3. Number words are, after all, generally named after the fingers and hands that are first used to represent quantities, both in the historical and ontogenetic sense of 'first.' Nevertheless, the eventual acquisition of words for numerical concepts greatly facilitates the mental manipulation of quantities, as kids learn to refer to precise cardinalities via lexical references. (This more precise lexical reference to quantities is verbal for most kids, but of course deaf children can also use linguistic signs to precisely and quickly refer to quantities.)

Through the acquisition of number words and counting, kids become successor-principle knowers and cardinal-principle knowers, and they also come to realize that large quantities are equal if they can be placed in one-to-one correspondence—they know what exact equality is. Note that, despite humans' innate number senses, none of these key principles is given to us instinctually. We have to work hard for them, over the course of many years in our childhood.

And we only do this hard work if we are spoon-fed by those around us the gifts of numbers and counting (sometimes as we are literally being spoon-fed). Those whose native cultures do not have (many) numbers or counting practices are not given the same tools to facilitate this work.[16]

Conclusion

So how exactly do we do the work of building on our simple innate capacities for numerical reasoning; how do we construct the edifice of distinctly human numerical thought? One influential and plausible account is that offered by Harvard psychologist Susan Carey. First, kids acquire counting words. Yet these are just words they learn as a memorized sequence—they do not recognize the precise association between 'two' and the quantity 2, for instance. The words essentially serve as placeholders for concepts that are to be filled in later. With time, and given enough exposure to such number words, they realize that some counting words have precise meanings associated with concepts that are easily distinguishable. Since they have a native capacity for distinguishing between sets of 1, 2, and 3 items, they have a natural basis for realizing what 'one,' 'two,' and 'three' represent, much as they have a natural basis for appreciating other linguistic distinctions such as plural vs. singular. They come to recognize that these words describe particular quantities and, given enough exposure, map the correct quantities onto those words. As noted above, first they become one-knowers, then two-knowers, and then three-knowers. They also come to infer that the other number words they learned in a sequence must correspond to other precise quantities as well. The sequencing of words, they realize, corresponds to a sequencing of quantities that differ by one, so that 'three' and 'four' have the same additive relationship as 'two and 'three.' The approximate number sense gives them a foundational awareness that higher quantities are distinguishable, so it likely plays some role in the process of acquiring other numbers. Eventually children come to realize 'five' means one more than 'four,' 'six' one more than 'five,' and so on. With enough familiarity and practice, they truly "get" the successor principle, the cardinal principle, the existence of exact equality for larger quantities, and so forth.[17]

In essence, children take realizations that they already have, say that 'three' is one more than 'two,' and build up other concepts by analogy, like 'six' is one more than 'five.' Words serve as guideposts in this process, letting children know there are precise numerical concepts that still must be generated. Learning numbers is a process not so much of labeling concepts, but of "concepting labels."[18] The labels, in this case sequential number words whose meanings kids do not fully grasp initially, serve as placeholders for the concepts that later fill them more adequately. This process of creating new concepts out of older ones to make sense of words whose meaning is not fully formed, is sometimes referred to as *conceptual bootstrapping*. Kids are pulling themselves up, conceptually, by their simple numerical bootstraps.

In its basic form this account has now been supported by much research, though primarily with children in large industrialized societies. The clear consensus, then, is that number words are essential to the developmental expansion of human quantitative reasoning beyond our limited innate reasoning. Number words are the key to unlocking the potential of our two number senses, or at least make it significantly easier to unlock this potential. Practice with number words and counting helps us to transfer exact quantity recognition to large quantities that we would recognize in a rough or erratic manner otherwise. This story is cogent, credible, and well supported experimentally. And note what a major role language plays in this account. Kids' understanding of numerical distinctions is a fantastic accomplishment, but it is one that depends largely on number words and counting practices.[19]

This discussion is inevitably truncated: we have not considered various stimulating findings related to the ways that children learn numbers and the quantitative skills associated therewith. Nevertheless, this discussion has stressed some of the crucial findings relevant to our larger purpose of illuminating the role that numbers play in our lives, and the role they have played over the course of human history. Numbers and counting routines transform how kids think about quantities. They provide a new level of precision to human numerical thought, one that is not simply a product of natural brain development. It is the result of growing up in particular cultures with developed traditions of counting and other related skills. These traditions and skills rest, ultimately, on number words.

7

QUANTITIES IN THE MINDS OF ANIMALS

Recently animal scientists have been sounding the depths of the intelligence of other species in novel ways. In laboratories and field sites around the world, researchers are showing that primates and many other types of animals are smarter than we might expect, or at least smarter than we once thought. Here is an example of one complex cognitive task used to test primate intelligence, a task implemented by scientists at the Max Planck Institute for Evolutionary Anthropology in Leipzig, Germany. A chimp is placed in a room with a transparent Plexiglas cylinder attached to a wall. This cylinder is narrow and fairly deep (5 centimeters wide, 26 centimeters long), making it impossible for the animal's fingers to reach its bottom. At the bottom of the cylinder is a peanut, placed there by a researcher. Chimps like peanuts. So the animal wants to eat the peanut but, maybe cruelly from its perspective, it cannot do so. Thankfully, there is a solution to this scenario, however unclear it may be initially. Located about 1 meter from the cylinder is a water dispenser from which the chimp can drink. This water dispenser is immovable, as is the cylinder. So what is a hungry chimp to do? It wants the peanut, but it cannot use its hands, nor does it have access to some long solid tool to try to stab the peanut. (Chimps in the wild are well known to use stabbing tools, for instance when killing bush babies for a little protein.) What the chimp does have, though, is a liquid tool. Water it can drink. When placed in this situation, many human children struggle for a solution. In fact, 4-year-old children given this task almost always fail—children only succeed fairly consistently around the age of 8. The solution, of course, is to transfer the liquid from the water dispenser to the cylinder, causing the peanut to float and rise to a reachable position in the cylinder. A significant ratio (roughly a fifth) of chimps are capable of realizing that they should use their mouths to get the peanut to float closer to the surface. And so they spit water over the peanut, one mouthful at a time. Some chimps are persistent enough that, after numerous trips, the peanut is floating higher in the cylinder and they can reach in and grab it. Success![1]

This is just one task that is indicative of the sort of cognitive dexterity evident in our closest genetic brethren. There are many others. Research with animals closely related to us, and with many animals not so closely related to us, is continually expanding our understanding of the cognitive capacity of those animals. From chimps to New Caledonian crows to whales, a series of findings in the past few decades has torn down many of the cognitive walls that we thought existed between *Homo sapiens* and other species. This research often uncovers some of the core mental capacities evident in the just-described experiment: Other animals are capable of planning. They are capable of tool use. They are capable of thinking through many novel problems in ways once thought impossible.

Some of these problems, it turns out, are arithmetic in nature. In this chapter we examine some of the quantitative capacities that other animals possess, considering some of the experimental work that has discovered those capacities. We devote much of our attention to research conducted with chimps and other primates, since these species are particularly relevant if we are to better understand how our uniquely human forms of quantitative thought evolved. From the outset, however, two caveats are in order. The first is that research on animal cognition, including animal numerical cognition, is constantly expanding. The history of such study is riddled with revisions, as researchers inevitably uncover new cognitive skills of which animal minds are capable. This will likely be the case, once again, in the coming years. The second caveat is that, when considering such research, care must be taken not to be too anthropocentric, nor too anthropomorphic, when interpreting results. This point requires expansion, as it is central to anthropology, primatology, and associated fields. The key point here is that we must let the data do the speaking and try to prevent any of our own natural biases from determining our interpretation of those data and what they might say about how animals think and behave. It is tempting to assume that animal cognition is irrelevant to research about humans, since humans are clearly a special kind of beast, with non-animal minds or perhaps infused with an ethereal spirit. Such a perspective can be termed an anthropocentric one. Depending on one's theological or theoretical proclivities, this can be an enticing position. But from an empirically grounded perspective, the evidence alone should dictate our views on animal thought. Rather than assuming that animals lack particular cognitive skills, skills that may be difficult to uncover given the limitations of cross-species communication, we must carefully rule out the existence of such skills.

Conversely, though, many people seem naturally given to an anthropomorphic viewpoint, according to which other animals are presumed to have many humanlike thoughts and emotions, since humans are "just another animal." There are reasonable motivations for suspecting the latter is not the case, however, and care must be taken not to project human characteristics onto animal thoughts and behaviors when the relevant data are equivocal. In my frequent interactions with college students, I find that the latter position is often prevalent, due in part to experiences with pets or other animals with which people often feel an intimate emotional bond. Perusals of Facebook, Reddit, or other social media invariably reveal videos or tales of pets that are ostensibly demonstrating their "love" of or "friendship" with their owners. And while it is indisputable that animals form attachments to humans (some anthropologists have suggested that animal-human relationships were essential to the evolution of culture), it is difficult to establish what is actually going on in the minds of other species. How much of their behavior, for instance, is due to their "feelings" toward us or each other, and how much is due to predictable stimulus-response associations? When animals do think and emote, are they thinking and feeling in ways that we, members of a species that has benefited from larger brains, culture, and language, would recognize as being anything like our own thoughts and emotions? Adequate answers to such questions are difficult to arrive at, despite any intuitions we might bring to the table. While we may have a strong personal inclination toward an anthropocentric or anthropomorphic view, that inclination is likely based on our own anecdotal experience. And the interpretation of personal anecdotes is ultimately an unsatisfactory basis for conclusions of a scientific nature, as evidenced by the fact that personal intuitions on such topics can vary dramatically. In other words, intuitions regarding the cognition of other species may say more about us than it does about them.[2]

In the context of animal numerical thought, an oft-cited yet still-important cautionary tale is that of Clever Hans. Clever Hans was a horse, a beautiful specimen of the Orlov Trotter breed. This horse belonged to a German named Wilhelm von Osten. Von Osten seems to have had an array of eclectic interests, which included teaching Hans mathematics and practicing phrenology, the now defunct area of study devoted to examining human skulls in an effort to explore the underlying brain modules supposedly dedicated to particular mental skills. One of his interests included demonstrating to the public, during the first decade of the twentieth century, that Hans was capable of an assortment of complex cognitive tasks. These tasks included reading and spelling German words, understanding a calendar, as well as finding solutions to all sorts of mathematical problems. For such tasks, Hans demonstrated his competency in the relevant mental skills by tapping out a sequence with his hoof. So, for example, if von Osten asked Hans to subtract 8 from 12, Hans would produce 4 taps. Many of the mathematical problems Hans appeared to solve were actually much more complex than this, and crowds across Germany were wowed by his intelligence as he answered most questions correctly. He became somewhat of a celebrity, and was reported on by publications as far away as the *New York Times*.[3]

Now, as you probably have surmised, Hans was not actually capable of performing mathematical operations, nor was he capable of understanding German. So, one might wonder, how was von Osten tricking the public by cueing Hans in ways audiences could not detect? Well, here is where the tale takes a perhaps unexpected turn: von Osten was not in fact duping his audiences (nor was he even charging them). In fact, when other people were allowed to ask Hans questions, the horse's performance did not substantially deteriorate. When these people did not know Hans or von Osten, Hans seemed to understand their questions and was able to offer generally correct responses. Enter Oscar Pfungst, a German psychologist who was much less enamored than the awed crowds of the equine spectacle. Pfungst was confident that there was some confound in the setup, some other variable that was enabling Hans to tap out the correct responses. Through a series of trials, Pfungst demonstrated that Hans was not in fact so clever at math. When Hans could not see the person asking a particular question, his tapping deteriorated to random responses. Furthermore, and this is a crucial point, when Hans could see the person but the person did not themselves know the answer to the questions they were asking, Hans's performance also deteriorated.

At least two conclusions can be drawn from this cautionary tale. First, although animals like Hans may not be good at math or reading, they are much better than we might realize at discerning visual information presented by humans. Even though the questioners were not consciously revealing answers to Hans, they were offering subtle visual cues that benefited his performance. Close inspection revealed that their heads moved slightly as the series of Hans's taps approached the correct answer. And somehow Hans was picking up on this unintentionally communicative cue. Second, we must beware of tendencies to anthropomorphize animal subjects. In the case of von Osten, for instance, he continued to tour with Hans after Pfungst's study of the animal, and remained unconvinced by the psychologist's results. Von Osten lost his impartiality, seemingly because he projected human characteristics onto Hans, maybe in part because of the social bond the two had developed.

The story of the Clever Hans effect is still circulated today because of its core lessons, which are as applicable now as they were more than a century ago. Consider the famous case of Koko the gorilla, which has been claimed to be able to speak "gorilla sign language" and thus communicate with humans. This has made Koko somewhat of a celebrity, and she has interacted with William Shatner, Robin Williams, Mr. Rogers, and various other famous personalities. However, many criticisms of the research of Francine Patterson, Koko's trainer, have surfaced over the years, given the strong social bond between Patterson and Koko. These criticisms have convincingly demonstrated that it is difficult to extrapolate from Patterson's interactions with Koko the actual range of the gorilla's communicative and cognitive skills. When strong social bonds develop between trainers and animals, trainers are unlikely to maintain complete objectivity and may tend toward anthropomorphic interpretations.

Accounting for the Clever Hans effect is not as straightforward as it may seem either. One might suggest, for instance, that trainers not be present when experiments are conducted, or that only doubleblind experiments be utilized. Such suggestions are not easy to implement, though, and in many cases their implementation may be impossible. After all, some experimental tasks require that animals obey the instructions of trainers they trust and that they have a social bond with outside of laboratory contexts. Removing the trainers from such contexts may quickly lead to the disintegration of the whole enterprise.

In addition to such concerns, nonhumans are obviously not linguistic, and so the methodological challenges discussed in Chapter 6, associated with studies of prelinguistic infants, also apply to animals. In some ways it is a wonder that we know anything at all about the mathematical cognition of animals. And it is no surprise that, even today, there are still some disputes over the extent of their numerical cognition. Nevertheless, despite these ongoing disputes and a host of methodological challenges associated with such work, we are beginning to get a sense of the numerical capacities of many species besides humans. While animals may not be capable of figuring out math problems like those given to Clever Hans, it turns out that their numerical cognition is not too dissimilar from that of human babies.

Numerical Cognition of Nonprimates

Regular quantity correspondences surface in animal behavior in unexpected ways. Consider this: in 1831, fur traders in Ontario reported that the Ojibwe indigenous population was facing a severe fur and food shortage, because one of their primary prey, the snowshoe hare, seemed to be disappearing. Not coincidentally, fur traders of the Hudson Bay Company reported an analogous shortage of lynxes, so prized for their soft pelts. Since lynxes also feast on hares, the shortage of the latter species seemed to have reduced the population of the former. The records of the Hudson Bay Company, dating back to the 1670s, reveal that such shortages of lynxes and hares co-occurred at regular 10-year intervals. Large-scale studies now suggest that these regular population shortages are due to predictable patterns in overpopulation. When a local ecology is saturated and cannot maintain any more hares due to their out-of-control breeding, this can lead to stressed food supplies and subsequent pronounced decreases in hare reproduction rates. These decreases have a knock-on effect on other species like lynxes, at fairly regular 10-year intervals.[4]

Or, consider the case of cicadas, a family of insects that has more than 2,000 representative species. One so-called periodical genus of cicadas spends most of its lifespan underground, feeding on the roots of trees. These insects only emerge from their underground existence to reproduce in very large numbers. After 2 months or so, during which they mate and lay eggs for the next generation, the adult cicadas are gone again. Depending on the population, adult cicadas will only reemerge 13 years later or 17 years later. This cycle is remarkably long and remarkably regular—it is almost as though the cicadas are counting the years until their reemergence. But of course this is not the case. What seems to have happened is that cicadas with such emergence cycles have been selected for by nature. Most animals that eat cicadas have 2–10 year reproduction cycles. Imagine if the cicadas emerged every 12 years, for example. If they did, they would be easier prey for all predatory species that have 2-year, 3-year, 4-year, and 6-year cycles, since 12 is divisible by 2, 3, 4, and 6. One would expect, then, that cicadas with 12-year cycles would face greater reproductive challenges. In contrast, cicadas with long reproduction cycles recurring in intervals like 13 and 17 years should face fewer challenges from other species. After all, these intervals are prime numbers, so they are not as easily divisible as, say, twelve. Environmental pressures have selected for cicadas with prime-number reproduction cycles and selected against those with more neatly divisible cycles.[5]

The cases of snowshoe hares and cicadas illustrate that regular quantities are evident in the behavior of nonhuman species, including insects. Yet they also demonstrate that, for such regularities to surface, other animals need not have working numerical cognition. In many cases we can assume they do not, given, for example, the inherent limitations of the nervous systems of insects. For example, we know that some ants have a mechanical recognition of the number of footsteps they have taken to reach a particular location, but this recognition does not definitively demonstrate that ants can conceptualize quantities.[6]

However, when we extend our discussion to more complex organisms, such as salamanders and various types of fish, we find that many species on distant branches of the tree of life have cognitive capacities for recognizing the differences between larger and smaller quantities. It is often unclear in such cases, however, whether these apparent quantity estimation abilities are due to confounds like the greater size, density, or movement associated with larger quantities. Consider the case of salamanders. In one study, researchers gave salamanders a choice between two and three tubes filled with a delicacy, fruit flies. The salamanders spontaneously chose the selection with more fruit flies. In a subsequent study, however, it was found that salamander selection of another delicacy, live crickets, was based on the amount of movement in the observed insects. When this movement was controlled for, the salamanders' selection seemed random with respect to the quantity of crickets they were viewing. In other words, salamanders are discriminating a continuous quantity of something (overall movement) when they make such choices, but they are not differentiating individual quantities of 2, 3, and so on. Though many studies in the wild have demonstrated that various species can form impressions of the greater amount of something, studies outside the laboratory cannot control for variables that are crucial to understanding how much of the amount-appreciation of other animals is simply due to their ability to recognize more or less "stuff" or movement in a continuous fashion, as opposed to their ability to discretely distinguish quantities.[7]

Rats—whom we do not typically think of as close relatives of ours but, as mammals, do share much of the human lineage—are in fact able to distinguish quantities. This point has been established for more than 4 decades now. In a study conducted in 1971, it was found that rats could be trained to approximate numbers. Rats that were rewarded for pressing a lever a certain number of times tended come close to that number after training. So, for example, a rat was rewarded for pushing a lever five times. When given an opportunity to push a lever later on, it did so about five times. The "about" here is crucial. It is not that rats are capable of exactly recalling that a lever should be pushed five times. They do have the capacity, though, to recall that it should be pushed approximately five times. In such cases, they are much more likely to push the lever five times than, say, eight times. But they are also more likely to push the lever four times than eight times. As the target quantity increases, so does the range of errors of the rat lever-presses. While the responses of the rats in the 1971 study were messy, they were normally distributed around the quantity they had been taught. Such noisy but frequently correct responses imply that rats, like many other species and like humans without numbers, are not capable of exactly differentiating most quantities. They are capable of approximation, though, and the fact that such a distantly related mammal is capable of quantity estimation suggests that the human approximate number sense was present in our common mammalian ancestor with rats. That species existed at least sixty million years ago.[8]

Some more distantly related species also share humans' ability for quantity estimation, though in such cases it is not clear that this is due to the shared inheritance of our innate numerical abilities. In other words, other animals may have *analogous* quantity recognition skills rather than *homologous* skills. Analogous features are similar features that exist across species because they have evolved independently to overcome similar environmental challenges. In contrast, homologous features refer to characteristics that exist in numerous species because of shared ancestry. For example, the four-leggedness of lions and bears is a homologous feature. In contrast, the wings of butterflies, bats, and birds are analogous features.

Some bird species, which are only distantly related to humans, possess the ability to estimate quantities, but it is uncertain how much this ability is due to analogous or homologous cognitive components when contrasted to humans. There are legends and stories of birds that could count precisely, but it is difficult to neutralize the apocryphal elements of such tales. Furthermore, many human owners are convinced of their pet birds' math skills (or those of other pet types for that matter). Yet such anecdotal reports do not factor into our discussion here, due to factors like the Clever Hans effect and our tendency to unjustifiably anthropomorphize our pets' emotional and cognitive states. Setting aside anecdotes, though, we also have strong experimental evidence that many nonprimate animals, including birds and rats, can approximate quantities. Yet even in experiments with animals considered relatively intelligent (compared to, say, salamanders), it is challenging to control for all variables to ensure that the approximation skill being uncovered is truly quantitative in nature.

Consider this example. When researchers presented lionesses in the Serengeti with audio of another single lioness roaring, the lionesses were likely to approach the source of the audio to fend off the fake intruder. In contrast, when the lionesses heard a recording of three other lionesses, they were less likely to approach the source of the audio. So can lionesses count the roars they are hearing? Maybe, but it is hard to tell from such a task. Perhaps they are just detecting the general amplitude of the roaring—they have some vague association between amount of roaring and amount of danger, without distinguishing in abstract numerical terms the quantity of either variable. Whatever the details, these perceptual abilities of the lionesses benefit their survival odds by letting them avoid unnecessary risks and suggest that they can likely discriminate the number of other lions they are hearing. In a similar vein, it is known that pigeons are able to consistently select the larger quantity of food items without training. Such a selection bias also confers clear benefits to the odds of an animal surviving and procreating. Whether avoiding danger or choosing more caloric food, estimating quantities helps animals excel in their varied environments.[9]

What seems clear from these and other findings is that many species are capable of discriminating quantities in approximate ways. Yet the results also hint that this quantity discrimination is sometimes based on a rough recognition of continuous variables (i.e., that some choices in experimental and nonexperimental settings reflect a preference for a greater "amount" of a particular thing). As cognitive scientist Christian Agrillo recently noted, "numerosity covaries with other physical attributes (i.e. cumulative surface area, brightness, density, or the overall space occupied by the sets) and organisms can use the relative magnitude of continuous variables to estimate which group is larger / smaller."[10] Distinguishing the amount of such associated attributes offers clear benefits to the survival and reproductive success of species. But this ability differs in nature from the specifically numerical capacities humans have at birth: the ability to discretely recognize and subitize small sets and the ability to approximate the number of items in larger sets.

Yet other experimental inquiries suggest more plainly that some nonprimates do share at least one of these specifically numerical capacities—the innate approximate number sense. Also, in the case of a few tested animals, for instance dogs or New Zealand robins, it seems that they can precisely distinguish the number of items in small sets, much like humans. In fact, New Zealand robins have been shown to discriminate one vs. two items, two vs. three items, and three vs. four items. Yet when contrasting sets larger than four items, they are only successful when the ratio between sets is at least 1:2, for instance if they are forced to choose between four and eight items. This aptitude for precise discrimination of smaller quantities and more approximate discrimination of larger ones is very reminiscent of what we observed vis-à-vis prelinguistic and anumeric humans.[11]

Surprisingly, some of the best intra-species evidence for distinct systems of number recognition comes from a species that is even more distant, phylogenetically, from humans. Recent work has established that guppies (a type of small fish) are capable of both discriminating small quantities and approximating large ones. Animal cognition researchers placed the guppies individually in an environment in which they could choose to join one of two visible groups of guppies. When each of the two groups exceeded four fish, the tested guppies selected the larger, safer group in most cases. Their selection benefited from greater ratios between the two groups, however. That is, they were more likely to choose the larger group if the discrepancy between it and the smaller group reflected a ratio such as 2:1, 3:1 or 4:1, with the likelihood increasing along with the size of the ratio between their options. The guppies also typically chose the larger of two groups when each group had four or fewer fish. Fascinatingly, though, this consistency was not impacted by ratio for smaller quantities. So, if one group had two fish and the other had four, they chose the latter group about two-thirds of the time. But if one group had three fish and the other had four, they still chose the latter group about two-thirds of the time. While humans are much more accurate on comparable tasks, there is an intriguing parallel in the responses of the fish. They, like us, seem to discriminate smaller quantities in a distinct manner when contrasted with how they discriminate larger quantities.[12]

There is a lot researchers have learned, yet much more left to be learned, about the numerical cognition of nonhuman primates. The picture so far depicted by the relevant work, of which I have described only a small fraction, remains a fuzzy one. In the case of many species, quantity discrimination appears to be based primarily on the perception of continuous variables like amount of perceived movement. Furthermore, the discrimination strategies used by some nonprimates might vary according to the task or the stimuli involved in the task, and future work will continue teasing out such variables. Such research will be crucial to our understanding of the evolution of neurobiological number systems, such as those present in humans. It is still unclear, for instance, how unique the numerical cognition of vertebrates is, or how distinct the innate numerical skills of primates are when contrasted to other smart vertebrates like birds. Furthermore, distinct lacunae remain in the experimental record of other animals. For instance, no work on the quantity recognition skills of reptiles has been systematically conducted. Filling in this gap in the experimental record will help us better understand how ancient the innate human numerical abilities are, and how much the numerical abilities of many species are due to homologous characteristics shared with humans. If similar abilities are observed in reptiles, we could have clearer evidence that some homologous numerical capacities extend back to the ancestral species of mammals, reptiles, birds, fish, and many other vertebrates—a species that lived more than 400 million years ago.

Numerical Cognition of Primates

The numerical cognition of nonhuman primates is particularly relevant to the story of numbers. Other primates, including apes like chimps, are our closest genetic relatives. This means their genome is very similar to the human genome. In the case of chimps, some research suggests there is an approximate correspondence of 99 percent of the genomes of our two species (as there is between our species and bonobos). We are similar in terms of our biology as a result of this correspondence, and so if we are concerned with better understanding our innate number senses, it is vital that we explore the minds of these animals and other related primates. Nevertheless, though we share a lot of our DNA-encoded genes with other primates, we still must take care not to unjustifiably anthropomorphize them. Our genetic overlap actually implies little vis-à-vis the numerical cognition of these cousins of ours.[13]

The characteristic double-helix structure of DNA molecules is due to the ladderlike bonding of only four nucleobases, referred to via the familiar symbols of A, C, G, and T (adenine, cytosine, guanine, and thymine, respectively). So there are four DNA components that ultimately make up genes, even the genes of very disparate species. Furthermore, many of these species have remarkable overlaps in their genetic material made out of DNA—their genomes. For example, there is about 25 percent overlap between the human genome and that of grapes. (And we have fewer genes than grapes!)[14] So some caution should be exercised before reading too much into percentages of genomic correspondence across species. I doubt, after all, that you consider yourself one-quarter grape. Nevertheless, our genome tends to be very similar to that of other mammals due to our common ancestry. For example, canine and bovine species generally exhibit about an 85 percent rate of genomic correspondence with humans. Given these factors, and given the remarkable behavioral disparities between dogs, cows, and humans, it follows that we should be careful when drawing any inferences based on the genetic overlap between chimps and humans. Certainly, we should not necessarily expect that chimps are capable of humanlike numerical reasoning just because they are closely related to us. After all, small changes in genetic makeup can, among other influences, lead to large changes in brain size. To understand the relationship of numerical thought in humans and their related species, then, we need to let the experimental data do the talking.

And the data are talking. For the past several decades, various intrepid researchers have been reconnoitering the cognitive worlds of chimps and other nonhuman primates, mapping out the numerical cognition of these animals. As a result of that reconnaissance, it is now clear that our primate relatives do in fact share some of our innate facility with quantities, and they also share some of the limitations that humans face in the absence of numbers. They have homologous numerical capacities that bear striking resemblance to our senses for both the exact recognition of small quantities and the approximate recognition of larger quantities.

In an experiment similar to some conducted with human children, psychologists found that rhesus monkeys are capable of telling apart small quantities. The monkeys were presented with different quantities of treats (apple slices), and these delectations were then hidden from view. The monkeys were then allowed to select which hidden quantity of treats they wanted. If their choice was between 1 and 2 treats, or 2 and 3 treats, or 1 and 3 treats, or even 3 and 4 treats, they consistently selected the larger quantity. In contrast, however, they proved incapable of consistently selecting the larger quantity of treats when the choice was between, say, 4 and 6 treats. In such cases their choices deteriorated to random selections, suggesting that their brains are wired to exactly differentiate only small quantities.[15]

In more abstract tasks, rhesus monkeys have also demonstrated that they can recognize discrepancies in larger quantities, but that this recognition is dependent on how disparate those quantities are. For instance, one study demonstrated that these monkeys can be trained to recognize arrays of items in ascending order. After being trained in this way, they learned to select arrays of 1, 2, 3, and 4 items sequentially. They were then presented with two groups of even larger quantities, and were capable of consistently touching the smaller quantity first—they could order the larger quantities just as they had learned to sequence the smaller ones. However, the speed with which they performed this task varied in accordance with the discrepancy between the larger quantities. The greater the discrepancy was, the faster the monkeys' reaction time became. In a follow-up study, the responses of two rhesus monkeys closely resembled the responses of eleven human adults tested, when the human subjects were prevented from using verbal counting. Such a resemblance is evidence of an ancient approximate number sense, inherited by both rhesus monkeys and humans.[16]

The species closest to humans on the tree of life, on the next branch over in fact, is also capable of discriminating quantities in fairly refined ways. Chimp quantity discrimination capacities are reminiscent of those observed in young humans. For instance, just as human toddlers tend to select the larger quantity of candies when presented with a choice, chimps tend to select the larger quantity of sweet treats when presented with two trays containing varied amounts of the treat in question. About 30 years ago, animal researchers observed that a chimpanzee given the option between two trays with small chocolate chips on them most frequently chose the tray with the greater number of chocolates. However, chimp performance became more chancelike for higher quantities when the ratio between the quantities of chocolates was not substantial enough. In other words, chimp selections were characterized by the same ratio effect observed in many other animals and in anumeric humans. More remarkably, the results in the relevant study suggested that chimps are not only capable of discriminating the larger quantity of chocolate chips given a sufficient discrepancy between the trays. They are also capable of adding quantities together before contrasting the overall amount of chips across the trays. For instance, in some cases the chimp was faced with a choice between two trays. On one tray were two piles of chocolate chips, one with three chips and the other with two. On the other tray were also two piles, one with four chips and one with three. In such cases, the chimp still generally recognized that the quantity of the first tray, 5 $(3+2)$, was less than the quantity of the second tray, 7 $(4+3)$, implying that chimps are capable of adding small quantities and comparing the results across sums. However, it should be stressed that the chimp selections were correct in the majority of cases only, and were riddled with errors. Furthermore, when the difference between the contrasted sums was small, say 8 vs. 7, accuracy disintegrated. While it is fair to conclude from such experiments that chimps are capable of spontaneously summing and contrasting quantities, it should be emphasized that their addition abilities are error prone, particularly when the difference between contrasted quantities is small. By now, this is a familiar pattern. Based on research like this and other experimental work not surveyed here, we can be confident that chimps have a natural ability to spontaneously approximate quantities and to exactly differentiate small quantities. And they are certainly not our only primate relatives to share some of our primal understandings of quantity differentiation.[17]

Research with primates has also shown that they are capable of learning numerals and associating these with ordinal and cardinal information. That is, they can learn to order symbols such as 2, 3, 4, and 5, while also appreciating that such a sequence corresponds to an increase in the set size of a given group of items—such items as snacks they might receive when demonstrating their understanding of the symbols. In fact, experimental work has shown that rhesus monkeys can learn to touch numerals 1–9 on a computer screen in ascending order, and that the monkeys can learn which quantities are represented by these numerals. This ability has since been demonstrated for squirrel monkeys and baboons also. Once they are trained properly with such symbols, squirrel monkeys seem to be able to add numerals together, as evidenced by their tendency to choose addends like $(3 + 3)$ over $(5 + 0)$, when given a pair of choices describing the quantity of treats they will then receive. Such choices are not completely regular, and the "monkey math" certainly contains errors, as we might expect. But the monkeys' choices are also clearly not random, and the tendencies in their choices reflect the power of symbols for refining quantity recognition. Numbers can be acquired by monkeys, then, though there are limitations to this acquisition not evident in humans.[18]

The mathematical abilities of nonhuman primates, like those of some other species, are clearly characterized by a distance effect and a magnitude effect. The former effect refers to the fact that these animals, like anumeric humans, are much better at recognizing quantity differences when the quantities being contrasted differ by a sizeable ratio. The latter effect refers to the fact that these animals discriminate smaller quantities more effectively than larger quantities. The cross-species prevalence of the distance and magnitude effects is one of the key findings that has been gleaned from work on this topic. This prevalence constitutes evidence for an ancient homologous approximate number sense and perhaps a homologous exact number sense as well, though much work remains to fully elucidate the innate numerical abilities of nonhuman species.[19]

Conclusion

Our innate numerical abilities are ancient and shared by many species to one extent or another. It is understandable that so many species would have ways to tell quantities apart, at least in approximate ways. Some judgments of quantities are essential to survival in the wild and therefore to the reproductive success that leads to the long-term preservation and proliferation of genetic traits. Whether one is talking about the benefits of rats or pigeons differentiating larger sets of food items from smaller ones, or the benefits of lionesses being able to distinguish larger sets of other lionesses, the survival advantages conferred by quantity recognition are obvious. Still, despite such intuitive understandings of why quantity recognition skills have been inherited across a variety of species, we do not know why these capacities have not been subsequently refined in most species.

In a way, the illumination of the cognitive abilities of other animals, particularly our big-brained ape cousins, brings further mysteries into the light. When we consider that some chimps can be trained to more accurately recognize quantities, this point becomes particularly salient. Why, if other species have the capacity to learn more elaborate kinds of numerical thought, have they not honed their own capacities over the millions of years they have been evolving on a separate branch of the tree of life? Chimps have the foundations of mathematical thought, but they never build much of anything on those foundations. The foundations evident in their thought, and in our own before we acquire numbers during development, seem pretty rudimentary. As Elizabeth Brannon and Joonkoo Park, two animal cognition specialists, recently suggested: "it is challenging to understand how such a primitive system that is not capable of representing exact large numbers could give rise to the formal mathematics that is uniquely human."[20] The quantitative thought with which we and other species are innately equipped is orders of magnitude removed from the kinds of quantitative thought that most humans eventually possess. This suggests that biological explanations of such thought are inherently limited. Most of our numerical cognition owes relatively little to our neurobiological equipment and owes much more to the ways we manipulate that equipment. This manipulation can only be possible if there are external tools interacting with our innate mechanisms for quantity differentiation. The principal external tools in question are numbers, symbolic representations of quantities that are linguistically reified and used in culturally variant ways. The existence of numbers explains the gulf between actual human numerical thought and the numerical thought to which we are innately predisposed.

Some of the evidence for the power of numbers also comes from other animals that have received extensive symbolic training in captivity. Perhaps the best example of such an animal is Alex, an African grey parrot trained for decades by psychologist Irene Pepperberg. While Alex died in 2007, results obtained on mathematical tasks conducted with the parrot were published as recently as 2012. Those results offer striking evidence that Alex was capable of performing arithmetic feats generally uncharacteristic of any other species besides *Homo sapiens*. In a series of experiments, Alex was shown to consistently label and order numerals, through verbalizations. More remarkably, he was found to label the quantity of sets of items, even if the sets included as many as eight items. Perhaps most astonishingly, Alex was able to add two sets of zero to six items together and was shown to arrive at the correct answer in most cases. Another nonhuman animal that has been shown in peer-reviewed research to sum two sets consistently is a chimpanzee named Sheba. Trained animal "geniuses" like Alex and Sheba appear to have precise mathematical skills enabling the exact differentiation of quantities larger than three. This is a startling finding given that nontrained animals, regardless of species and brain size, do not exhibit this skill. But notice what is required before such animals show themselves to be geniuses: years and even decades of learning, during which the animal trainers familiarize their trainees with symbols for quantities. The trainers teach them numbers. In some cases, and certainly in the case of Alex, the animals manage to learn these numbers. Much as children eventually learn to symbolically encode quantities beyond three, some trained animals may do so as well.[21]

This is a remarkable observation made possible through such research: the human invention of numbers can be applied across species, at least in some cases. As Pepperberg observes when discussing animals like Alex and Sheba, only those animals "trained to represent quantities symbolically with Arabic and / or vocal numerals . . . appear to exactly map such numerals to precise cardinal values of sets."[22] So, while chimps and parrots are capable of forming abstract concepts of precise higher quantities, the human invention of numbers is what makes such abstractions possible.

In the last three chapters we have observed that anumeric human adults, prelinguistic human infants, and various animal species can think about quantities in approximate ways. They can also think about them in precise ways when they are considering small quantities. These approximate and precise abilities serve as an essential foundation for the construction of more elaborate thought about quantities. Yet it is a rough foundation. Construction on this foundation requires the usage of symbolic tools. It requires the use of numbers—verbal and written symbols for quantities. In Part 3 of the book we explore how numbers were likely invented and examine the profound ways they impacted the human experience.

PART 3

Numbers and the Shaping of Our Lives

8

INVENTING NUMBERS AND ARITHMETIC

Patterns in language yield patterns in thought. Extensive research has now demonstrated that differences between languages can yield differences, often subtle ones, in the cognitive habits of their speakers. This finding, commonly referred to as *linguistic relativity,* has now been supported by dozens of studies on topics like spatial awareness, the perception of time, and the categorization of colors. For instance, as we saw in Chapter 1, "where" the future and past "are" depends on the language you speak. Similarly, the manner in which you recall and discriminate colors is affected in subtle ways by the basic color term inventory of your native language. Our tour of numberless worlds ultimately led us to the conclusion that numeric language also yields differences in how people think. Number words, present in the vast majority of the world's languages (though not all of them), certainly influence quantitative cognition. Only those people who are familiar with number words and counting can exactly differentiate most quantities. The presence of numbers in a language does not just subtly influence how we think about certain quantities, then; it also opens up a door to the world of arithmetic and mathematics. The first step through that door is the realization that quantities, regardless of size, can be precisely differentiated.[1]

But how exactly do numbers first open this door? And what happens after we walk through it? In Part 3 of the book we address such questions. In this chapter we consider the "how"—how do counting words and basic arithmetic come into existence? I present an account of how humans likely invented (and still invent) basic number words, and how we take those number words and use them as building blocks in basic arithmetical processes.

Unnatural Numbers

The findings from numberless worlds suggest plainly that we need numbers to really "get" quantities in ways that are uniquely human. But, as I mentioned back in Chapter 1, this raises a paradox. If we need numbers to appreciate most quantities precisely, how did we get numbers in the first place? How could we ever name the amounts in particular sets of items, if we could not recognize the amounts? If we could not recognize, for example, that 'seven apples' represents not 'six apples' and not 'eight apples,' how did we first come to use words like 'six,' 'seven,' and 'eight'?

Given the apparent intractability of this paradox, some have concluded that humans must be innately predisposed to acquire number concepts. According to this perspective, we must be wired in such a way that we can differentiate 5, 6, 7, and so on during the course of our natural cognitive development. Yet, prima facie, such an approach is problematic. If we are predisposed to recognize different set sizes as separate abstract entities, then what is the limit of this predisposition? Are we naturally predisposed, for example, to eventually realize that 1,023 is not 1,024? This seems fairly implausible. Framed differently, nativist views on numbers just delay the point at which we reach the paradox.

In his generally superlative book on language and numbers, linguist James Hurford noted that number words are names for the "non-linguistic entities denoted by numbers."[2] That is, number words label conceptual entities. In a related vein, archaeologist Karenleigh Overmann recently suggested that "quantity concepts must surely precede their lexical labels, or there would be nothing to name . . . a method of invention cannot presuppose that which it invents."[3] This latter stance is understandable, but it arguably trivializes the extensive evidence we have reviewed in Chapters 5–7. According to that evidence, words for quantities beyond three do not simply label preexisting concepts, because those concepts do not exist for most people until they actually learn numbers.

In my view, this is the key to resolving this paradox: words for quantities beyond three make concrete the precise numerical abstractions that are only occasionally and inconsistently made by some people. Some of these people may eventually invent numbers, but if they do not, their fleeting abstractions are not transferred to others. The naming of such ephemeral realizations is what eventually enables people to consistently recognize quantity distinctions. This notion of consistency is crucial to resolving this quandary, I believe. It seems that humans, as a group, only inconsistently show the ability to make a simple but powerful realization, the realization that sets of quantities greater than three can be identified precisely. This simple realization has led, in all likelihood more times than could be documented, to the invention of symbols for such larger quantities. These symbols are chiefly verbal in nature, judging from the fact that the overwhelming majority of the world's cultures have words for such quantities though most cultures traditionally lack written numerals or elaborate tally systems. Some people invented number words to concretize the potentially transient recognition of the existence of exact higher quantities.

Does this mean that number words simply serve as labels for concepts? Not really. The truth seems a bit more nuanced than the forced dichotomous choice assumed by this paradox. Number words are not simply labels, yet they do describe conceptual realizations that *some* people make *some* times. The term "label" implies that the words simply denote concepts that we all think about: concepts all humans are born ready to appreciate (at least eventually), regardless of their cultural environment. But clearly not all humans have such concepts at the ready even as adults, and likely most people would never make the relevant realizations that can be described via numbers. Just as clearly, though, some people have made those realizations, even if inconsistently. In those real historical cases in which people managed to describe that realization with a word, they invented numbers. The concept they named was subsequently recognized by other members of their culture through the adoption of the relevant word(s). Number words are conceptual tools that get passed around with ease, tools that most people want to borrow.

The general account I am suggesting is not really that radical. In fact, it could be similarly applied to countless human realizations that are described with new words. Words are frequently developed or invented to refer to newly uncovered concepts and realizations, not to innate ones. Consider an example like "light bulb." Numerous inventors in the late nineteenth century recognized that passing electricity through a metal filament resulted in incandescence, and various patents for short-life light bulbs were filed. Thomas Edison and his employees refined the art of encasing a filament in a vacuum glass bulb that would allow it to burn for an exponentially longer duration. In a sense, the invention of this kind of light source was based on a simple realization—keep the wire out of contact from surrounding air, and it will remain bright for much longer. This simple realization was not hard for others to grasp, and the resultant device was no doubt easy to name. Certainly the compound word "light bulb" is not hard to decipher. Nevertheless, despite an inherent elegant simplicity, and despite the ease with which the concept and associated term "light bulb" can be acquired by most people, nobody would seriously maintain that people are innately predisposed to understand light bulbs. "Light bulb" may describe a particular concept that is not too hard to understand, but it is not a natural one. Number words also serve as references to simple realizations. We may not be innately pre-equipped to make those realizations, but some humans do make them, and other humans can acquire them through linguistic means. As noted in the Prologue, what makes our species so special is not so much that we are great at inventing, but that we are exceptional at inheriting and sharing inventions because of our linguistic nature. (Edison himself once noted that he was "quite more of a sponge than an inventor.") So we are not naturally predisposed to have concepts like "light bulb" floating around in our head, waiting to be labeled. Nor do we have concepts like 6, 7, and 8, hovering in our minds, awaiting labels. To invent numbers we must first recognize, in an experientially haphazard manner that is dependent on erratic realizations made in some minds only, that the namable exact quantities (beyond three) exist.

There is evidence that grammatical number, for instance the distinction between plural and singular, has a different origin when contrasted to lexical numbers like 'six' and 'seven.' As noted in Chapter 4, grammatical number distinctions in the world's languages discriminate 1, 2 (less commonly), and 3 (rarely), from each other and from all greater quantities. Such distinctions *do* seem to refer to preexisting concepts, given that our brains are innately equipped to discriminate such quantities. The same could be said of small number words like 'one,' 'two,' and 'three,' which are labels of natural concepts. These concepts have a clear neurobiological basis and the words labeling those concepts are therefore common to the vast majority of the world's languages. Not coincidentally, they often have different historical sources than larger number words in a given language. As psychologist Stanislas Dehaene notes, "oneness, twoness, and threeness are perceptual qualities that our brain computes effortlessly without counting."[4] But other number words do not have the same neurological basis and are not computed so effortlessly. Language and other symbolic facets of culture afford *Homo sapiens* the opportunity to invent such numbers. Exploiting this possibility is clearly not a straightforward matter, however, as evidenced by the fact that cultures vary extensively in the complexity and bases of their larger number words. As discussed in Chapter 3, though, this cross-linguistic variation is not random and hints at some clear underlying tendencies. Those tendencies suggest plainly that, while humans can take different routes to invent symbols for the quantity correspondences they occasionally recognize, those symbols are usually invented via the hands. This manual route is crucial for numbers greater than four.

In a recent study of the history of number words in Australia, linguists Kevin Zhou and Claire Bowern observed an interesting pattern consistent with the general account being discussed here, according to which numbers for 1–3 have a more basic status than larger ones. They found that the number word for 4 in these languages is frequently compositional, based on smaller numbers. In Chapter 3, I noted that this same pattern is also observed in the Amazonian language Jarawara, since the word for 'four' in that language is *famafama,* or 'two two.' Analogous patterns are found in many other languages in Amazonia, Australia, and elsewhere. The common compositionality of 'four' suggests it is a less easily named concept, a concept that humans often label by appealing to simpler, preformed ideas. Yet, even though we can and do name 'four' compositionally through the combination of smaller numbers, this is an anomalous number. Numbers greater than 'four' are not usually created via the simple composition of more basic quantities like 2, but instead by naming quantities after the hands somehow. While Australian languages can lose or gain number words with time, Zhou and Bowern concluded that those languages tend to acquire numbers greater than 5 relatively rapidly after the introduction of a word for 'five'—which is etymologically related to 'hand' in the most widespread Australian language family. This sort of finding, combined with the well-known quinary and decimal bases of the majority of the world's number systems, suggests that a manually based number 'five' serves as the foundation to a more productive sort of enumeration. It is frequently a gateway to new forms of numerical thought.[5]

The body serves as the basis of most number words beyond 'four.' People learn that their fingers can be placed into one-to-one correspondence with small sets being counted. Finger counting helps motivate the fact that words for hand(s) so frequently serve as a historical source of number words and bases. For all the merits of this account of higher number words, though, it glosses over some important questions: How and why does the correspondence between fingers and set sizes become established initially? If anumeric humans—whether prelinguistic children or Nicaraguan homesigners or the Munduruku—struggle with exactly recognizing higher quantities, how did any adult human 'number inventors' come to recognize the exact correspondence of fingers, specifically, with other countable items in sets as large as five or ten? Why are the fingers really so special in the progression toward elaborate numerical thought?[6]

There are at least two reasons, I believe. First, the fingers are special, because they are the most experientially basic discrete units in our lives. As was mentioned in Chapter 6, we become acquainted with our fingers in utero. And babies are manually obsessed, sucking on their fingers early in life, and visually focusing on their hands when they realize these items can be manipulated and drawn in and out of view under their own control. So fingers are incredibly salient all our lives. Crucially, they are not so salient in those related species that share similar innate capacities for numerical cognition. Consider gorillas, gibbons, chimps, and other ape relatives of ours. Because they are not bipedal like us, all of these species often use their forelimbs for locomotive purposes. We, however, use our hands largely for tool making and tool usage. We use our fingers for a greater range of purposes, including more specialized purposes requiring refined dexterity and manual focus. We are more finger focused than any other species, and this point may represent the first part of the answer as to how we sylleptically took hold of numbers with our hands.

Yet, despite the salience of fingers in our day-to-day lives, mere experiential ubiquity is unlikely the only motivator for the realization some number inventors make when they recognize that sets of fingers can correspond to sets of other items. A second reason fingers play such a prominent role, one that has not (to my knowledge) been discussed in the literature, is that they naturally occur in symmetrical correspondences. Not only are we constantly interacting with and focusing on our fingers but, since each finger has a natural similar-looking partner on the opposite hand, the potential for fingers to correspond to other items in a one-to-one manner is consistently underscored in our visual and tactile experience. The symmetry of our hands and fingers, and our frequent exposure to that symmetry in our oddly manual lives, likely has pushed many people to recognize the potential for associating sets of five items with each other.

Some people may also just come to realize the correspondence of a set of five body-external objects with a set of five fingers (with or without first labeling the equinumerosity of the fingers on each hand). The one-to-one alignment of fingers with such objects likely benefits from the natural physical alignment of small handheld items with the fingers, perhaps as those items are laid out on the palm of a hand. Whatever route number inventors take, however, the invention of elaborate number systems typically relies on the quantitative equality between fingers and specific items, including other fingers. Throughout human history, people have discovered such regular correspondences between quantities and then seized those quantitative correspondences—often by naming a particular correspondence after the word for 'hand' in their language. And when they did name it, they invented a key symbolic tool that facilitated subsequent reference to, and recognition of, the specific quantity involved and allowed it to be transmitted to the minds of others.

Adopters of new number words may then generate greater number words. Perhaps by combining the word for 'two' and 'hand' they innovate a word for 'two hand' ('seven') when counting items with their fingers. With time, practice with such a word would lead to its greater productivity, and it could be used in many contexts to name sets of different items. Other speakers might extend the innovation in new directions, creating words for 'ten' and 'twenty,' likely also basing such words on the number of digits on the human body. Many routes could be taken as number words are amassed, borrowed, adapted, and extended. Given the utility of numbers, and given the sorts of processes they enable, in most contexts they would spread in a population and potentially across populations in contact. They could occur in new languages as loanwords or as calques, in which the concepts are adopted cross-culturally but new words are made up for the borrowed concepts.

The realization that fingers can correspond to each other symmetrically and to other objects in a one-to-one fashion takes us beyond the capacity of our innate number senses. But it is a capriciously formed realization, judging from the fact that some languages do not have robust number systems and other languages have numbers that are not based on the hands or fingers. Nevertheless, this realization is the most prevalent factor at play in the verbal reification of quantities greater than 4. (In a sense, it is remarkable that not all groups of people name larger quantities after their hands, given the anatomical biases we have toward equating the five fingers on one hand with the five of the other.) Ultimately, the discovery of the existence of large precise quantities, and therefore the invention of most numbers, is an accidental by-product of our bipedalism, like many other distinctly human things. Bipedalism eventually yielded a greater manual fixation and the recognition of the symmetry of our fingers, and it also facilitated the occasional recognition of the one-to-one correspondence of fingers with other countable objects. As a result of such factors, our hands offered the path of least resistance in our trek toward numbers.[7]

The phenomenon at play here—humans making sense of quantities because of features of their brain-external bodies—is an instance of the larger phenomenon known as *embodied cognition*. Over the past few decades, many philosophers, psychologists, linguists, and others have pointed out that a host of human cognitive processes are based on, or at least facilitated by, features of the human physical experience. The burgeoning study of embodied cognition has demonstrated that a variety of thought processes are constrained or augmented in accordance with our anatomy and the functioning of our bodies. While this is not the place for a lengthy discussion of such processes, consider again the example of temporal perception discussed back in Chapter 1. We, as English speakers, think of the future as being in front of us, because, in a way, we walk into the future. As we walk, the moments in our past occurred while we were located in physical spaces that are now behind us. Conversely, future moments are expected to occur while our bodies are located in spaces now ahead of us. The 'future is forward' metaphor is the result of thinking of time in terms of this physical and corporeal experience. This common perspective on time is an example of embodied thought, since our construal of time's progression is impacted by the way our bodies work. In a similar vein, when humans think 'five objects are like a hand' or some related thought, and subsequently invent numbers, features of their bodies are enabling the cognitive process in question. Their cognition is embodied as they refer to a hand, metonymically, to refer to a quantitative characteristic of that corporeal feature. While linguists have long recognized that number systems tend to be decimal, quinary, or vigesimal (or some combination thereof), the extent to which embodied quantitative thought impacts the creation of most numbers has been underappreciated in many circles.[8]

Beyond Simple Counting

Many number-based concepts are not manually inspired. Elaborate mathematical practice has only been developed independently in relatively few cultures, despite the universality of fingers and despite the worldwide prevalence of number words. Hand-based numbers do not even necessarily yield extremely large numbers. Just because a language has words like 'five' and 'ten' does not imply it has words like one 'thousand' or one 'million.' Simple number words are a necessary condition, but not a sufficient one, for the elaboration of more complex numbers.

Many other kinds of numbers also do not necessarily follow the introduction of basic number words in a given culture. The anatomical pathway leads to simple integers that have been termed "prototypical counting numbers"[9] like 'five,' 'six,' 'ten,' and 'twenty.' But it does not necessarily yield a number like 'zero.' (See the discussion of zero in Chapter 9.) Nor does it necessarily yield fractions. Nor negative numbers. Nor irrational numbers. Nor the discovery of the Fibonacci sequence. And so on. A natural question in the course of this discussion, then, is how we begin to build on the prototypical counting numbers to arrive in a world with all sorts of other numbers and mathematical concepts. A complete detailing of the evolution of arithmetic is outside the scope of this book. Yet it is worth considering some of key factors that are involved when humans take prototypical counting numbers and use them to innovate core mathematical concepts, such as addition, subtraction, and multiplication. After all, the latter concepts are pivotal to many human material and behavioral technologies.

This discussion of the expansion of numerical concepts draws us back to a recurring theme: humans make sense of abstract notions through concrete things found in our physically grounded lives. Much as metaphors for time and the manual route toward numbers are based on how are bodies are structured and work, a convincing case has been made that basic arithmetical cognition is grounded in our physical experience. That is, the evolution of mathematical concepts out of prototypical counting numbers results in large part from humans appealing to metaphorical thought centered on physical entities.[10]

Two key kinds of physically grounded metaphorical thought are relevant here. The first is simply conceptual metaphor, of which there are numerous cases in any given language. To cite two additional examples from English: the abstract notion of an emotional mood is depicted in terms of temperature, and negative things are described as being 'down.' The former physically grounded metaphor surfaces, for example, when we talk about people having 'warm' or 'cold' personalities, with an implicit assumption that 'warm' things are somehow more friendly or inviting. The latter physically grounded metaphor is evident when we speak about feeling 'down,' or of somebody feeling like they are 'down in the dumps,' or that they 'fell into depression.' These metaphors are of course just two of many. (The metaphorical connection between things being physically 'down' and emotionally sad stems from experiential associations such as that between death / burial and being 'down' in the ground.)

The second sort of metaphorical thought relevant here is what linguists refer to as *fictive motion,* wherein we think of items as moving while we mentally scan them. For example, if I say "the skyscrapers of the Miami skyline run alongside Biscayne Bay," this is a case of fictive motion. Obviously the skyscrapers are not actually in motion, but I speak of them as though they are. Similarly, if I claim that "the border of Peru and Brazil runs through Amazonia," nobody interprets this to mean that the border is moving. Physically grounded metaphors and fictive motion are both central to the construction of mathematical reasoning.

Cognitive scientists, most prominently Rafael Núñez of the University of California, San Diego, have offered evidence that basic metaphors and fictive motion help structure arithmetic. They note, for instance, that in the creation of addition and subtraction practices, one figurative linchpin that structures the relationship between numbers is the 'arithmetic is object collection' metaphor.[11] In other words, people think of numbers in terms of objects, once again turning abstract notions into something more concrete and tangible. Some evidence for this metaphorical orientation is the overlap between terms and phrases for addition / subtraction and the manipulation of physical entities. I can speak of "adding two and five," or I can speak of "adding cheese to my burger," or "adding salt to my salad," or "adding another piece of furniture to the room." I can say "the addition of three and three equals six," or I can say "that new Ferrari is a great addition to his car collection." And, much as I can speak of "combining three and five," I can talk of "combining sugar, eggs, and butter." We bring together numbers in our minds like we bring together objects in the brain-external world. Conversely, we separate numbers like we separate objects. I can say that "if you take away that pillar, the structure will crumble," or I can say thanks for "taking away the trash." But I can also speak of "taking five from seven," or say "twelve take away six equals six." Furthermore, just as I might say "fifteen minus two is thirteen," I can say "minus the belt, that outfit does not work." I could go on, but the point is clear enough: a lot of the language we use for collecting and removing objects is also used for collecting and removing numbers.

The object-number linguistic parallels do not end there, however. Just as we can speak of the size of objects, we can speak of the 'size' of numbers. I can say that a trillion is a "really big" number, or that "seven is smaller than fifteen." I might say, "I don't know exactly how much she makes, but I know it's a huge number." Or I might claim that "her salary is tiny compared to what she deserves." We frequently talk about numbers as though they are manipulable objects of various sizes, objects that can be compared, combined, and collected. This sort of metaphorical thought is so natural that we may not realize that we use it. Some of this physical grounding of arithmetical language may be because the invention of numbers relies so heavily on our physical bodies. But it is also because, in many cognitive domains, we think and speak of abstract notions as though they were elements in the physical world. This metaphorical transfer enables us to more adroitly handle such abstract concepts. Thinking of numbers in terms of physical objects facilitates their mental storage, representation, and manipulation, since we can visualize and recall objects more easily than we can abstract concepts.

Fictive motion also plays a role in the development of arithmetical strategies, though likely a more minor one. This is ultimately a distinct but also metaphorical phenomenon, and the metaphor at play might be termed "arithmetic is motion along a path."[12] The basic idea is this. Many people (certainly English speakers) describe numbers as occurring in a line and describe movement along this line. Linguistic instantiations of this metaphor are plentiful. For example, I can say that "101 and 102 are very close." If I ask you what ten plus ten is and you say thirty, then I can claim that your answer is "pretty far off." If you see a bunch of people in a classroom, you might state that there are "around twenty" students there. If you do not think there are quite that many, you might state that there are "nearly twenty." Children might be told to count from one to a hundred, without "skipping any numbers." Or they can learn to "count backward" as though counting entails movement forward or backward along a number line. The naturalness of such language can obscure what is happening when we use it—we are talking about abstract numbers and the quantities they describe as though they exist on a line that that can be moved along or scanned. Reference to numbers in terms of lines and manipulable objects is pervasive and facilitates our acquisition of many arithmetical concepts during childhood. These metaphors are employed deliberately in pedagogical contexts as well, as students associate numbers with physical objects and with number lines in mathematical workbooks.[13]

In addition to patterns in speech, patterns in co-speech gestures reveal the ways we conceptualize numbers metaphorically. The study of the gestures people make while speaking is a fertile area of inquiry in cognitive science, with much research suggesting that gestures serve as a window into human cognitive processes. For example, the fact that English speakers tend to think of the future as being in front of their bodies is reflected in their pointing forward while talking about future events. Conversely, they often point behind themselves when talking about the past. Similarly, gestures surface when people are talking about numbers, as evidenced in a recent study of video recordings of college students. When the recorded students spoke of adding numbers together, they simultaneously used 'collecting' gestures or 'path' gestures. For the latter gestures, they moved their fingers / hands from one side of their body to the other, as though numbers were progressing along a line. The former gestures involved the inward movement of the hands, with the hands configured as though the students were grasping or holding something. As the students were speaking of adding numbers, then, they were simultaneously and nonconsciously collecting fictitious objects together with their hands, or moving their hands along a fictitious line. Metaphors clearly play some role in constructing mathematics out of numbers, though the extent of this role requires further exploration.[14]

The conclusion that humans often think of numbers in terms of physical space is supported by other sorts of findings as well. For example, people are quicker to make mathematical judgments when spatial and numerical information correspond neatly. Consider, for instance, which of the following two numbers has a greater value:

7   or   9

Or which of the following two numbers has a greater value:

6   or   **8**

Did you take longer to evaluate the first pair of numbers? If so, your reaction times are like those of most people who have made such judgments, since spatial and numerical information are somewhat tricky to completely disentangle. In fact, their entanglement also surfaces in brain imaging research. When people are asked to focus on a number or conduct numerical reasoning tasks, certain portions of their brain are activated. Similarly, when they conduct tasks requiring judgments of physical size and / or location, those same portions are activated.[15]

The spatial-numerical overlap is also evident in the spatial numerical association of response codes, commonly referred to as the SNARC effect. (The discovery of this effect, like many other features of numerical cognition, owes itself to French psychologist Stanislas Dehaene.) The SNARC effect surfaces in experimental contexts when subjects are asked, for example, to press a button as soon as they see a particular number on a screen. For larger numbers, the subjects respond faster when they press the button with their right hand. For smaller numbers, they respond faster when pressing with their left. This suggests that numbers are thought of as being in a spatial configuration along a line, with smaller numbers to the left and larger numbers to the right. However, in some cultures with right-to-left writing, the line appears to be reversed, and larger numbers are responded to more quickly with the left hand. The commonness of the SNARC effect, like the spatially influenced numerical judgments mentioned in the preceding paragraph, is indicative of the cognitive intertwinement of space and number.[16]

Although the evidence for some metaphorical and spatial basis of arithmetical thought is convincing, this does not imply that metaphorical reasoning is the only basis of such thought. In fact, it is implausible that any one single factor explains the evolution of numbers or the manipulation of those numbers through basic arithmetic. In the case of metaphors, for example, we find that how cultures frame numbers in terms of space can vary. Furthermore, there is little to no evidence in some cultures that people place numbers on a number line. (See the discussion of the Yupno in Chapter 5.)

At a more basic level, we have already observed that a few languages lack precise numbers altogether or have a limited number set. In addition, while the manual route to numbers is so common, it is not the only possible route, and the bases of number systems do not always show vestiges of this path. For instance, some languages use senary (base-6) systems that cannot be ascribed to the usage of words for hands or fingers in the initial naming of numbers. (See Chapter 3.) One recurring theme of linguistic fieldwork is that we should not overgeneralize patterns evident in many languages and assume that they exist in all languages. However, given that all people share the same basic brains and bodies, it is not surprising that they usually take similar paths to arithmetical concepts. And these paths often involve metaphorical routes.

Another basis for the expansion of numerical thought is more specifically linguistic. Work by linguist Heike Wiese suggests that syntax, the way in which sentences are structured, facilitates the creation of numerical concepts. Linguistic structure may help us transform number terms like 'five' and 'six' into more productive numerical systems. Eventually we, as number users, come to realize that 6 is one more than 5, and one less than 7—we come to appreciate the successor principle. But we likely make this realization, at least in part, because language offers us practice with symbols whose meaning varies in accordance with the sequence in which they occur. Consider a transitive sentence such as "The crocodile ate the snake." The meaning of this sentence depends on its individual words, but the words alone are insufficient for comprehension in the absence of syntactic convention. After all, some snakes (anacondas) eat crocodiles, so how do we know who is eating whom? The ease with which we interpret such potentially ambiguous sentences is due to English syntax. Because subjects generally precede verbs, which in turn precede objects, we know that the snake is the one that was eaten. If we extend this sort of sequence-dependent meaning to the world of numbers, we can see how syntax helps us appreciate the relationship of counted numbers to one another. We may be predisposed toward creating and deciphering counting sequences, and appreciating that count words have specific predictable meanings, because linguistic structure helps lay the groundwork for understanding number sequences. From this perspective, syntax gives us the basis for realizing that word meanings can vary in accordance with the order in which they occur.[17]

Such factors as metaphor and syntax are useful in understanding how we, as a species, went from coming to grips with basic numbers (pun intended) to manipulating them in new ways. They help explain how, out of prototypical counting numbers, arithmetical processes evolved. To be clear, I am not suggesting that these factors occur necessarily in every culture, or that they follow logically in all cases from the invention of number words. It would obviously be inaccurate to state that "people invent numbers because of their fingers, they have metaphors and language, and, yada yada yada, they eventually arrive at the fundamental theorem of calculus." The extent to which cultures exploit basic arithmetic and more intricate mathematics varies dramatically, suggesting that many socially contingent factors are at work. Nevertheless, the factors we have discussed are apparently common and key components in the refinement of mathematical thought.

Numbers in the Brain

The human brain is several times larger than it should be, given the typical ratio of brain size to body size that is evident in other primates. About 80 percent of our brain's mass is found in the cerebral cortex, that messily folded layer of gray matter, about 2–4 millimeters thick, that is divided into two hemispheres and four major lobes. The cortex has, according to some estimates, 21–26 billion neurons that enable all sorts of uniquely human forms of thought. Extensive research in the past few decades has explored our massive brains and examined, among many other topics, how numerical cognition happens neurophysiologically. Brain imaging studies have found the place in the cortex where most numerical thought happens. This numerical locus is not in our uniquely overdeveloped frontal cortex. Instead, much of our basic numerical reasoning takes place in a region called the intraparietal sulcus (IPS), first mentioned in Chapter 4. As was noted then, our innate number senses seem to be housed primarily in the IPS.[18]

Given that human innate number senses are not so dissimilar from those of other primates, it may not be surprising that much of our numerical thought occurs in a region of the brain that also exists in the brains of closely related species. In fact, brain-imaging work suggests that the IPS of monkeys is also activated in response to numerical tasks, such as figuring out whether two sets of dots represent different quantities. Furthermore, particular sets of neurons in their IPS activate in accordance with the specific quantities focused on in particular tasks. When one object is perceived by monkeys, a predictable group of neurons fires. When two objects are perceived, a different but still predictable set fires.[19]

Studies of our own brains have shown time and again that the IPS of both hemispheres fires during many numerical processing tasks. Brain imaging studies often involve functional magnetic resonance imaging (fMRI). In such studies, subjects conduct numerical reasoning tasks while their brain activity is monitored via fMRI. As some neuroscientists have noted, this activity is largely localized to a specific portion of the IPS, a horizontal strip of the cerebral groove referred to as the horizontal IPS (hIPS). A variety of imaging experiments have demonstrated that the hIPS lights up when humans perceive and discriminate quantities. For example, if you were presented with an array of dots, or asked to quantitatively contrast two arrays, your hIPS would activate. Fascinatingly, the hIPS fires when quantities of dots are perceived, when numerals are viewed, or when spoken numbers are heard. In other words, numerical processing across different modalities happens there. The hIPS is associated with abstract numerical thought, not simply with the visual perception of groups of objects. Furthermore, the degree to which it is activated corresponds to the intensity of numerical thought required for a particular task. For example, if you were asked to judge whether twenty dots were greater in number than five dots, your hIPS would activate only weakly. In contrast, if you were asked to discriminate twenty dots from seventeen dots, the degree of activation would be much higher.[20]

The neural imaging data paint a similar picture to that depicted by other data we have looked at: humans have a basic, primordial neurobiological component that is shared across numerous species. But we also have the capacity to expand the functions of this component beyond the realm of differentiating small quantities and approximating larger ones. This functional expansion requires the use of other parts of the human cortex. More specifically, we need to use portions of the left hemisphere associated with linguistic processing to expand numerical thought into the realm of exact discrimination, exact addition, exact subtraction, and so on. For the expansion to happen, we need a way to verbalize quantitative distinctions. This verbally based expansion, which is apparently facilitated by linguistic syntax and metaphors, is evident in imaging data that shows the activation of language-associated regions of the cortex during some quantitative tasks. The brain imaging data lead us back to the same familiar conclusion, then. To build on our innate mathematical thought, we need verbalized symbols for quantities. We need numbers.[21]

Conclusion

This is how the invention of numbers has most frequently occurred: people realized, inconsistently, that precise quantities such as 'five' exist. Such realizations led, in some cases at least, to the creation of words for those quantities. The words used have typically been based on already existing names of body parts, body parts that enabled or facilitated the relevant realizations that precise quantities exist. The resultant number words represent quantities exactly, and this exact representation stems partially from our innate capacity for basic quantity appreciation. However, the pivotal role of our fingers and hands in less basic quantity appreciation cannot be overstated. This role is due in some measure to the ubiquity of fingers in the human cognitive and perceptual experience, and to the inherent symmetry between a person's hands. It is due, indirectly, to bipedalism. And it represents a key way, but one of many, in which humans make sense of their cognitive experience through embodied thought.

The invention of basic numbers, prototypical counting words, is only the beginning of the tale. The usage of such words leads to the eventual functional expansion of neurophysiological activities associated with quantitative reasoning. While we do not completely understand this expansion, we know that it is largely dependent on the existence of verbal numbers. Other linguistic phenomena, including metaphors and regular syntactic sequencing, help construct the edifice of arithmetic, but this edifice is grounded on verbal numbers.

Numbers are a contrivance of the human mind whose effects on the story of humanity have been profound. They transformed our understanding of quantities. But their effects were not just cognitive, as they have also come to mold our experience in other ways. Next we discuss the extent to which numbers shaped, and shape, other facets of our everyday lives.

9

NUMBERS AND CULTURE:
SUBSISTENCE AND SYMBOLISM

At the summit of Khufu, the greatest of the pyramids on the Giza plateau, the mathematical wizardry of ancient Egyptians is literally laid bare. Millions of tons of weathered limestone converge on the megaliths at the square apex, a space barely large enough for an adult to relax after reaching the summit, surrounded by precipitous drop-offs on all four sides. (Though the cigarette butts scattered on the summit suggest many people have somehow been able to chill out up here.) Looking down the sides of the pyramid from the top, swallowing any acrophobic tendencies, one is struck by the geometric regularity of the concentric squares of stone leading toward the base. For 139 vertical meters, the stacked squares incrementally grow larger as they approach the ground. In contrast to the miniscule top square, the bottom square has sides that are about 230 meters long. Interestingly, the perimeter of the base of the pyramid has a length that is almost exactly two times the original height of the pyramid when that height is multiplied by π.[1] There is still some disagreement as to whether this correspondence is intentional, or whether it is simply the coincidental by-product of the overall symmetry and precision of the structure.

What is indisputable is that this pyramid represents a remarkable and timeless achievement of ancient architects and laborers. It stood as the highest human-made structure for nearly 4 millennia, until the completion of the Lincoln Cathedral in England in 1311. The pyramid has served as a tomb for Egyptian pharaoh Khufu since its construction was finalized circa 2540 BCE—the stunning structure is so old that it has been a tourist attraction for thousands of years now. Its construction, like that of so many other great monuments, tombs, and buildings, required the dedicated labor of significant portions of the population that created it. It became a centerpiece of their culture, reifying the mythological and reinforcing various spiritual values. The construction of Khufu and other pyramids helped shape Egyptian life. The pyramid even plays a prominent role in the lives of many contemporary Egyptians profiting from tourism, 4.5 millennia after it was built.

It does not take much reflection to realize that the structure, like so many other human achievements of the past few thousand years, would have been impossible without numbers and mathematics. Large projects formed by collaborative cultures, which in turn help shape those cultures in a material-social feedback loop, are generally contingent on mathematics. Nobody would dispute that numbers enabled certain obvious features of material culture, like Khufu and other enormous structures, to bloom. In this chapter we explore a few of the ways in which numbers shifted the daily experience of humans in most cultures. In addition to such cases as Khufu, there are less obvious but more pervasive aspects of material and symbolic culture that have been created through numbers or have been enabled because of the existence of numbers. Mathematics has long been acknowledged as a crucial precursor to architecture, to industrialization, to the advent of medical and scientific procedures, and so on. For all its merit, though, this common acknowledgment arguably reflects our historical myopia, as we generally focus on the relatively recent developments in Western mathematics that yielded subsequent advances in engineering and science. Here the focus is on a more distant time-depth, and on the more foundational role that numbers had—apart from their role in advanced mathematics—in the construction of cultural practices that reshaped the human experience. Most pivotally, perhaps, numbers made possible agricultural revolutions and the associated innovation of writing systems in different parts of the globe at different moments in history. Elaborate mathematical practices developed in Mesopotamia, China, and in Central America only after agricultural revolutions enabled food surpluses that allowed for mathematically trained classes of people in those regions. Prior to such developments, though, agriculture and writing themselves were preceded in antiquity, not coincidentally, by the advent of number systems. In a very real sense, such number systems were the foundation on which large civilizations, including the civilization that created the pyramid of Khufu, were built. The claim here is not that elaborate number systems always yield expansive agricultural states or writing systems. The claim is that the usage of such systems was a necessary criterion for the dawning of such phenomena. Next I offer some evidence in support of this view.

Numbers and Subsistence

In ways that are often subliminal, we filter our experience through the learned and shared beliefs, values, and tools that we inherit from those around us. That is, we filter our experience through our native culture. Nearly every element of our lives is affected, in one way or another, by this filtering. What does it mean to be married? Well it depends on your native culture—in some it means having several spouses, in others it means having one. Maybe with a lifetime contract, maybe not. In my own experiences in disparate cultures I have witnessed marriages between two twelve-year-olds, between an eleven-year-old female and a forty-year-old male, between one man and several women, between two women in their fifties, along with myriad other marital arrangements. In each of the cultures in which I met the people in these marriages, their arrangements seemed to them natural, though they are obviously considered highly inappropriate in some other of the world's societies. These arrangements reflect some crucial variations in the notion of childhood across the world's cultures as well. This is not the place for a litany of the ways culture constructs our social lives, but if it defines what is meant by childhood and marriage, few aspects of our lives fall outside its purview. As I have previously noted, the linguistic component of a culture can also impact everything from how we think about space and time to the discrimination of color categories, sometimes in persistent (if subtle) ways. In short, our cognitive and behavioral experience is in many ways a product of culture-specific factors. These factors include the numbers used by particular groups of people.

As we have seen, number terms influence basic numerical thought, enabling basic quantity recognition skills. These skills can then be built on in the creation of arithmetic. Furthermore, it appears that basic quantity recognition skills and arithmetic enable or at least facilitate associated cultural changes, for instance, those related to subsistence type. In turn, those changes may place pressures on the members of particular cultures to further develop their inventory of number words and to refine arithmetical strategies. There appears, in other words, to be a symbiotic relationship between numbers and nonlinguistic facets of culture, as they act on each other over the course of generations. Some of the evidence for that symbiotic relationship is uncovered by examining global associations between behavioral components of culture and numerical language.

One of the most fascinating things about numbers (from a cross-cultural perspective anyhow) is the extent to which they vary. This variation is clear from the findings surveyed in this book. It is worth noting that most facets of meaning do not vary so much vis-à-vis their range of expression across the world's languages. For instance, in the case of color terms, there is in fact variation with respect to how the world's languages encode hues. But this variation is restricted in terms of the number of basic color terms, and languages generally have between three and eleven basic color terms. Similarly, terms for basic emotions and even smells have been shown to vary in analogous ways, as have a variety of other categories of concepts. Yet when it comes to quantity concepts, languages vary more dramatically with respect to how many words they employ—in order of magnitude, at an exponential scale. This is not to suggest that there is not some regularity in the way that number expressions are structured, or that number terms' meanings are not constrained. Although the amount of number words across languages varies substantially, due to the limitless extent of the quantities that can be named, the translatability of these words is usually pretty straightforward. The word for 6 in any language refers to precisely six items, by definition. This makes sense, as physical objects come in discontinuous quantities, in discrete sets. In contrast, colors of the visible light spectrum bleed into one another. As a result of this and other factors, the physical reference of terms for colors often differs a bit from language to language (see Chapter 1), as those languages divide up visible light in constrained but disparate ways. In contrast, the meanings of words like 'five' across languages are generally constant.

Still, given the unparalleled extent to which the *range* of spoken numbers can vary cross-culturally, it is particularly crucial that we explore the relationship between number systems and other cultural factors. The exploration of the culture-number nexus has been undertaken by various researchers over the years. Here we discuss one of the most crucial results of that work—the demonstration of a relationship between number system types and subsistence strategies. Recent work has offered evidence for a clearly discernible correlation between simple number systems (sometimes bordering on the nonexistent) and hunting-gathering subsistence. In contrast, agricultural modes of living are associated with more elaborate number types.[2]

Led by Patience Epps of the University of Texas, a team of linguists recently documented the complexity of the number systems in many of the world's languages. In particular, the researchers were concerned with the languages' upper numerical limit—the highest quantity with a specific name. This limit is not easy to establish in the case of some languages. This is true, for example, of Bardi. This Australian language has words for 1–3, but the status of the word for 4 is less clear, since it entails the reduplication or doubling of the word for 2. (As was noted in Chapter 8, the number for 'four' in Australian languages is often composed of smaller numbers.) Furthermore, the word *ni-marla* or 'hand' may refer to 5, but only for some Bardi speakers. Such an idiosyncratic number exemplifies the occasional challenges of establishing upper limits in particular number systems. In most cases, though, ascertaining the largest specific number in a given language is a more straightforward enterprise. The linguistic team in question found the upper numerical limits in 193 languages of hunter-gatherer cultures in Australia, Amazonia, Africa, and North America. Additionally, they examined the upper limits of 204 languages spoken by agriculturists and pastoralists in these regions. They discovered that the languages of hunter-gatherer groups generally have low upper limits. This is particularly true in Australia and Amazonia, the regions with so-called pure hunter-gatherer subsistence strategies. The limited numbers common to languages in these regions was mentioned in Chapter 3, but now we can relate those limitations to cultural factors.[3]

In the case of the Australian languages, the study in question observed that more than 80 percent are limited numerically, with the highest quantity represented in such cases being only 3 or 4.[4] Only one Australian language, Gamilaraay, was found to have an upper limit above 10, and its highest number is for 20. Given that all indigenous Australian populations traditionally rely on hunting and gathering, the association between limited number terms and subsistence type leans heavily in the predicted direction on that continent. The association is also robust in South America and Amazonia more specifically. The languages of hunter-gatherer cultures in this region generally have upper limits below ten. Only one surveyed language of hunter-gatherers in South America, Huaorani, has numbers for quantities greater than 20. Approximately two-thirds of the languages of such groups in the region have upper limits of 5 or less, while one-third have an upper limit of 10. Similarly, about two-thirds of African hunter-gatherer languages have upper limits of 10 or less. These ratios of limited number systems are much more pronounced than what we might expect from a random sample of cultures. In short, the correlation between basic subsistence strategies and numerical complexity is pervasive across these regions.

For all their usefulness, one issue with such findings is that they are based on a simplistic categorization of cultures into hunter-gatherer vs. non-hunter-gatherer types. Such a categorization is necessary to conduct such survey work, but it is important to bear in mind that human subsistence strategies vary more radically than such labels imply. For instance, so-called hunter-gatherer groups differ widely in the extent to which they acquire calories through hunting and in the kinds of hunting that they do. After all, hunter-gatherer groups in Australia, Amazonia, and elsewhere, hunt

different kinds of animals and live in distinct ecologies. These ecological variances include differential access to freshwater sources
and, therefore, varying rates of reliance on fish and other aquatic
animals as a source of calories. Furthermore, many hunter-gatherer
groups rely at least somewhat on slash-and-burn growing strategies, even if they do not do any sedentary farming. Finally, the kinds
of larger social networks in which particular hunter-gatherer groups
are embedded differ dramatically. Some hunter-gatherer groups in
Amazonia live fairly untouched existences. In fact, according to new
satellite imaging techniques, there are several isolated groups of indigenes in the region. In contrast, most hunter-gatherer groups in
the Great Basin region of the American Southwest (a region focused
upon by Epps and colleagues) have long been more interconnected
to one another and to larger communities. With such interconnectedness comes a greater prevalence of trade and the valuation of
commodities. Such a prevalence yields a heightened utility of
number words, while greater social interconnectedness leads to an
enhanced likelihood of number-word borrowing. In sum, different
hunter-gatherer cultures face markedly diverse socio-cultural pressures for heightened number usage. The homogenizing usage of
terms like "hunter-gatherer," while understandable, obfuscates
some of the important distinctions across culture types. For that
reason, it is not particularly surprising that hunter-gatherer groups
in North America generally have more elaborate number systems
than hunter-gatherer groups in Amazonia, for example.[5]

Despite the limitations of the terminology used to categorize
human populations, there is still a clear correlation between subsistence strategy types and the complexity of number systems: populations that rely on hunting and gathering, with little use of
farming or other complex agricultural techniques, are quite likely
to rely on modest numerical technologies. I should emphasize the
"quite likely" here, as there are exceptions—societies without ag-

riculture but with relatively high number limits. A few exceptional societies have some basic agriculture but low number limits. (Like the Munduruku, discussed in Chapter 5.) However, there are no large agricultural states without elaborate number systems, now or in recorded history. Still, I should be clear that the account being offered is not a deterministic one, according to which number systems with higher limits lead inevitably to agriculture. Instead I am suggesting that robust number systems are (and were) an important factor that help(ed) create agricultural practices. But ultimately the claim is one of coevolution between number systems and patterns of subsistence, as certain kinds of subsistence (like sedentary agriculture) also create pressures for the development of more elaborate number types.

This conclusion has ramifications not just for our awareness of modern peoples but also for our understanding of humans throughout history. After all, for the vast majority of our species' existence, we lived as hunters and gatherers in Africa, without elaborate agricultural practices and without intricate trade networks. A reasonable interpretation of the contemporary distribution of cultural and number-system types, then, is that humans did not rely on complex number systems for the bulk of their history. We can also reasonably conclude that transitions to larger, more sedentary, and more trade-based cultures helped pressure various groups to develop more involved numerical technologies. In fact, this transition was evident in our discussion of the genesis of writing in Chapter 2. Written numerals, and writing more generally, were developed first in the Fertile Crescent after the agricultural revolution began there. As people in that region developed large farms, and as life in the region became more reliant on agriculture, new pressures to precisely record quantities of commodities were placed on the peoples in that part of the world. They faced pressures to enumerate stores of wheat, barley, and millet, for example, and to record the number of these and other traded commodities resulting from agriculture and / or manufacturing in farming-dependent urban centers. These pressures ultimately resulted in numerals and other written symbols, such as the clay-token-based numerals discussed in Chapter 2. The numerals then enabled new forms of agriculture and trade that required the exact discrimination and representation of quantities. The ancient Mesopotamian case is suggestive, then, of the motivation for the present-day correlation between subsistence and number types: larger agricultural and trade-based economies require numerical elaboration to function. Like written numerals, higher number limits make agriculture and trade work, since they allow for the exact discrimination of all relevant quantities.

Even some supposedly simple agricultural practices are impossible without prior elaboration of number systems. From all the evidence we have surveyed, it is safe to say that the agricultural revolution would not have happened if people did not have extensive sets of numbers. As we have seen, people need numbers to discriminate most quantities precisely. So number words and other numeric tools are obviously required for us to keep track of the lunar cycle, astronomical cycles more generally, and other basic environmental features that are essential to the development of many agricultural practices. In large part because of this essentialness, many early numerical records in disparate regions like Mesoamerica and Mesopotamia track astronomical cycles and seasons. Without elaborate calendars based on numbers, humans could not track subtle celestial patterns, such as the recurring positions of the sun at different times of the year. We invent numeric tools, and these tools can then be used in unforeseen ways and refined to meet unforeseen needs. The use of numbers enabled people to keep track of the recurrence of the spring equinox and the winter solstice, for example, and this proved pivotal to agriculture. Number systems with high limits also enabled Sumerians and others to tally rows of barley in precise ways, or to precisely measure stores of grain for the winter. Without numbers, such tasks are not just difficult but simply impossible. We can now state this with confidence because of the recent experimental research discussed in this book. So numbers make agriculture possible and, ultimately, agriculture yields more sedentary, larger societies. These larger societies then yield bigger networks of minds sharing the same language, through which new numerical tools can quickly diffuse.

This diffusion of numerical tools has not been characteristic of the past alone, for instance during the agricultural revolution in the Fertile Crescent. In fact, the spread of number systems continues to change modes of living and likely is as rampant today as it has been at any point in the human story. Various sorts of socio-cultural pressures act on individuals in our own era, forcing them to adopt and adapt number systems. Consider the case of one of my good friends, a member of an indigenous group known as the Karitiâna. For anonymity's sake I will refer to this friend as Paulo. The Karitiâna population consists of about 350 indigenes living in southwest Amazonia. Most of these people live on a reservation about 90 kilometers by road from the growing Brazilian city of Porto Velho. But a growing percentage of the Karitiâna also live in Porto Velho. Paulo spent the majority of his childhood, in the 1980s and 1990s, in the largest village of his people's reservation. In a sense, he was raised on a jungle island, surrounded by Brazilian roads and farms. During that time, he learned Karitiâna numbers (see Chapter 3) but was also exposed to Portuguese numbers and written Portuguese numerals. The same could be said for the remaining Karitiâna of his generation. While some Karitiâna sought to make a living in nearby Porto Velho, many strived to maintain their traditional way of life on their reservation. At the time this was feasible, and their traditional subsistence strategies of hunting, gathering, and horticulture could be realistically practiced. Recently, however, maintaining their conventional way of life has become a less tenable proposition. A large hydroelectric dam is now in operation nearby, and this has impacted local fisheries. Similarly, the Karitiâna lands are being encroached on more directly by a growing population of local Brazilians, and this encroachment has yielded lower quantities of game and fish on the Karitiâna reservation. In short, for all the beauty and usefulness of their jungle island, many Karitiâna feel they have little choice but to seek employment in the local Brazilian economy if they are to survive. This is certainly true of Paulo. He has been enrolled in Brazilian schools for some time, has received some higher education, and is currently employed by a governmental organization. To do these things, of course, Paulo had to learn Portuguese grammar and writing. And he had to learn numbers and math also. In short, the socioeconomic pressures he has felt to acquire the numbers of another culture are intense.

More to the point, Paulo is but one of hundreds of millions of speakers of endangered languages today who face similar pressures to acquire others' numbers. By some estimates, more than 90 percent of the approximately 7,000 languages in existence today are endangered to one extent or another. They are endangered primarily because people like Paulo are being conscripted into larger nation-states, gaining fluency in more economically viable languages. These languages, typically European in heritage, have elaborate numerical and mathematical systems, particularly when contrasted to the languages of small populations that rely on hunting, gathering, and horticultural practices. From New Guinea to Australia to Amazonia and elsewhere, the mathematizing of people is happening. To survive and / or strive, indigenes are continually forced into greater contact with hegemonic nation-states. Prolonged interaction with the cultures of these nation-states typically requires the acquisition of complex number systems. This point is illustrative of the pattern that has existed, at a more region-specific level, for millennia: cultures place strong pressures on one another to adopt numbers and other numerical technologies. Just as numbers enabled cultural shifts, like a greater reliance on agriculture, cultural shifts also enable the acquisition of new numbers. Behavioral culture and number systems operate synergistically and in a feedback loop. Changes in cultural practices often require the acquisition of new numerical tools, which in turn facilitate new cultural practices, which, once again, may require further elaboration of numerical tools. And so on.

The Overlooked Advantages of
Some Number Systems

In speaking of the potential advantages (like agriculture) afforded by robust number systems, I should be clear that I am not equating the adoption of such systems with the "evolution" of culture or language. Many linguists and anthropologists in the nineteenth and early twentieth centuries developed an unfortunate habit of equating the features of European languages and cultures with some supposedly later stage in the evolution of human societies. The same could be said of European colonists, who tended to construe Western culture as the zenith of human social adaptation. Such views have long since fallen out of favor, at least in part because extensive fieldwork has shown them to be garbage. For example, many settlers once considered the languages of indigenous Americans to be primitive, lacking the grammatical refinement of modern or classical European tongues. Although some nonlinguists may still hold such outdated views, early-twentieth-century anthropological linguists like Franz Boas and Edward Sapir long ago dispelled them in academic circles. They showed how so-called primitive indigenous languages are filled with all manner of grammatical complexities not evident in Indo-European languages. Which is not to imply that indigenous languages are more complex either. For some time the consensus in linguistics has been that there is no way to objectively rank languages in terms of their inherent complexity. Furthermore, since it is generally agreed that all languages can be traced back to some African origin, we have no basis for concluding that some are more evolved than others.

Given these points, it is important not to fall into the same trap again. It is clear that languages vary in terms of numerical elaboration, in that some have much larger inventories of numbers. This does not imply, however, that the languages in question are more complex overall, or that the speakers of those languages are somehow more advanced socio-culturally. It does imply, though, that speakers of those languages have tools at their disposal that facilitate certain kinds of behavior. That much is uncontestable. Nevertheless, this does not signify that cultures can be located on some simple path to modernity, nor that all people should be concerned with such a path. The case of the Karitiâna underscores how involuntary the adoption of number systems is for many people. Intriguingly, even in the face of such pressures for adoption, some groups remain uninterested in utilizing a large set of numbers. This is certainly the case among the Pirahã, who are generally resistant to most aspects of Brazilian culture. Does this make Pirahã culture less advanced? If one employs a Eurocentric view of what it means to be advanced, certainly. Yet the Pirahã seem generally content with the choices they have made to maintain their culture, and their own ethnocentrism belies the simplistic conclusion that they themselves feel inferior or primitive when contrasted to outsiders. Their proud cultural lineage has long been well adapted to their environment, surviving in Amazonia for millennia.[6]

The traditional Eurocentric perspective toward linguistic complexity, which by default viewed indigenous tongues as primitive, did not just result in oversimplifications of the grammars of other languages. It also resulted in glossing over the numerical complexities in some languages. It was frequently presumed, for example, that number systems were necessarily less complex if they were not decimal in nature, or if they were utilized by cultures without writing systems. Now scholars have begun to realize that some indigenous number systems offer specific advantages for particular mathematical tasks—advantages that are not afforded by European number types. This appreciation of esoteric numbers has been spearheaded by the work of two cognitive scientists, Andrea Bender and Sieghard Beller. In the past decade they have published fascinating studies on the heretofore unrecognized cognitive advantages of the number systems native to some Pacific islands. Their studies suggest that the intricacies and beauties of some indigenous numerical systems can sometimes be overlooked.[7]

In Chapter 3 it was mentioned that, while most spoken number systems are based on the human body, there are exceptions. Some senary (base-6) numbers in New Guinea, for example, seem to have arisen because of common patterns in arrangements used for yam storage. Such number types are often perceived as simple by outsiders, because they are most effective in particular contexts and cannot be as easily abstracted to all countable items. In some languages numbers or number-like words may in fact be restricted to specific contexts. For example, Balinese and other languages, including some in South Australia, use birth-order names. These are not numbers, but they are personal names based on the order in which siblings were born. If someone is named *Ketut* in Balinese, for instance, we know that they were the fourth child born in their family. In the Australian language Kaurna, the first-born through eighth-born children can be distinguished by name endings.[8] Such number-like words are used in naming only, so the terms are not abstract enough to be considered proper numbers. Yet the conclusion that number systems with object-specific counting terms are less abstract and less productive is poorly grounded in some cases. In fact, the research of Bender and Beller on Polynesian languages demonstrates that some number systems with object-specific counting terms can present clear cognitive advantages to their speakers.

Consider the case of the language once spoken on the French Polynesian island of Mangareva. That language employed different counting sequences depending on whether speakers were enumerating, for example, breadfruit, pandanus (palmlike trees), or octopi. This remarkable variation could be considered primitive or unevolved by outsiders, since there was no single set of abstract number terms that could be used to count all items. Interestingly, though, the Mangarevan language descended from Proto-Oceanic, a language that apparently had a uniform decimal system that could be used to count anything. So Mangarevan counting, as well other related counting systems in Polynesia, developed out of a decimal system that some might think of as being more evolved than the Mangarevan one. Such a development contravenes the assumption that Mangarevan numbers represent some early stage in the development of true numerical complexity. The counterintuitive historical trajectory of this number system is likely due to the fact that object-specific number systems may increase the speed of mental arithmetic in some contexts while also reducing the cognitive effort of some mathematical tasks in the absence of writing.

Mangarevan had one main number system underlying the more overt number terms used to count specific kinds of objects. This main system was in fact a decimal one. Overlaid on this decimal framework, though, the language also had other counting sequences for more effectively counting specific sorts of objects. These sequences were interrelated in nuanced ways. The word *tauga,* for example, signified 1, 2, 4, or 8 items, depending on the kind of object that was being counted. There is a clear binary step-up from one type of *tauga* to another, since $2 \times 2 = 4$, and $4 \times 2 = 8$. But the decimal quality of Proto-Oceanic also remained in Mangarevan counting, since *taugas* could be grouped by tens. That is, Mangarevans counted the quantity of *taugas,* and in a decimal manner. So the word *paua,* for instance, signified 20, 40, or 80, depending on the item being counted. In other words, *paua* referred to 10, but to ten *tauga* values associated with a particular object—it is these latter values that could vary in a binary fashion. So *paua* meant, in essence, 10×2, 10×4, or 10×8, depending on what was being counted.

Essentially, Mangarevans were counting objects, mainly those important to their culture and trade networks, by pairs, quadruples, or eights. As Bender and Beller suggest, if someone counted twelve *taugas* of fish, they would be referring to 24 fish. If they counted twelve *taugas* of coconuts, they would be referring to 48 coconuts.[9] Mangarevans were counting objects not as individual items but as easily discernible groups. This counting-by-groups strategy would present advantages for items that come in predictable quantities of 2, 4, and 8. Analogous processes are at work today when we count items that naturally occur in groups. If you ask for somebody to buy you drinks at the store, for instance, you may ask for "four six-packs" rather than "twenty-four beers." The Mangarevan counting system was specialized for quantities that were typically encountered in the local ecology.

Additionally, the Mangarevan system hints at the possible advantages of a binary strategy for quickly grouping quantities, since the size of the quantity referred to by *tauga* was based on the power of two. Leibniz famously showed the advantages of binary-based computations back in the first part of the eighteenth century. Bender and Beller's research suggests that Mangarevans had seized on some of these advantages centuries prior to his work. The seemingly primitive and nonabstract counting native to some Pacific islands turns out to be not so primitive after all. Such a finding serves as a cautionary tale. Some superficially "unevolved" number systems function effectively, and in deceptively complex ways, to meet the needs of those who use(d) them.

Recent research also suggests that the complexity of some non-linguistic number systems have been underappreciated. Many counting boards and abaci that have been used, and are still in use across the world's cultures, present clear advantages to those using them. Few would doubt this conclusion if they observe people using, for instance, the Soroban abacus of Japan (derived centuries ago from the Suan Pan abacus of China). Children in Western industrialized societies generally lack familiarity with abaci, and they may seem primitive compared to the calculators common to classrooms in much of the world. However, unlike such calculators, the abacus presents some cognitive advantages. That is because, research now suggests, children who are raised using the abacus develop a "mental abacus" with time. That is, they internalize the structure of the abacus, using a mental image of the abacus to perform calculations via the imaginative manipulation of beads. According to recent cross-cultural findings, practitioners of abacus-based mathematical strategies outperform those unfamiliar with such strategies, at least in some mathematical tasks. The use of the Soroban abacus has, not coincidentally, now been adopted in many schools throughout Asia. The effectiveness of the mental abacus suggests once again that non-Western numeric symbols present some clear advantages, vis-à-vis those most of us are more familiar with. This effectiveness also underscores another key point: numeric technologies provide us with new ways of mentally manipulating quantities, ways that might be unpredictable prior to the invention or adoption of the technologies in question. This is true whether these technologies are new counting words, new abaci, or some other symbolic representation of quantities.[10]

The Influential but Languid Journey of Zero

Piercing the Cambodian jungle's canopy, a series of immense camouflaged sandstone faces stare out of the vast sculpted complex that is the Temple of Bayon. The complex lies in the ancient Khmer capital city known as Angkor Thom. The dozens of meters-high faces located around the complex resemble a blend of the visages of bodhisattva Avalokiteshvara, who is said to embody Buddhist compassion, and Khmer King Jayavarman VII. (See Figure 9.1.) The exquisite complex was built at the behest of Jayavarman VII, a purportedly beneficent figure who ruled these jungles some 900 years ago. History recalls Jayavarman VII kindly, in part because he established more than 100 hospitals to care for the citizens of his empire. Just several kilometers from Angkor Thom is Angkor Wat, the world's largest religious structure, built during the reign of Jayavarman VII's father. Like the great pyramids of Giza or those of Mesoamerica, the temples of the Khmer empire hold a special place on our collective imaginations. At Angkor, separated in terms of time and space from Western civilization, citizens of that empire constructed some of the world's most dazzling structures. They also created networks of hospitals and roadways, and established systems of irrigation without parallel. They did so with remarkable precision, and the symmetry and artistry evident in the remnants throughout Angkor inspires awe.

Amid all the physically imminent structures of the Khmer empire, it is easy to overlook some of the key technologies that were at the heart of this fabulous human accomplishment. Those technologies are not readily apparent as one walks around the

9.1.   One of the faces of Bayon, Cambodia. Photograph by the author.

stone facades and courtyards of Bayon, but the latter are ultimately vestiges of the technologies in question. You can probably
guess the technologies to which I am referring: spoken numbers
and written numerals. Furthermore, one particularly remarkable
and innovative numeral seems to have facilitated the construction
of the Khmer empire, having arrived from the Indian subconti-
nent just centuries before the faces of Bayon came to life. The nu-

meral in question is 'zero.' This circular symbol for nothingness is still evident in contemporary Cambodian culture, for instance on Cambodian currency. Although it is tempting to see a circular symbol for zero in Cambodia today and assume it is derived from the Western 0, the reverse route of influence is more accurate. In fact, in 2015 the world's oldest known unambiguous inscription of a circular zero was rediscovered in Cambodia. The zero in question, really a large dot, serves as a placeholder in the ancient Khmer numeral for 605. It is inscribed on a stone tablet, dating to 683 CE, that was found only kilometers from the faces of Bayon and the other ruins of Angkor Wat and Angkor Thom. As noted in Chapter 2, the Maya also developed a written form for zero, and the Inca encoded the concept in their Quipu. Similarly, there is some trace of the concept in Babylonian inscriptions. Yet the zero we all know and love, the round symbol for nothingness that facilitates so many mathematical operations, was not utilized by the Greeks, Romans, or most other ancient civilizations. In fact, it was not really used at all in the Old World until it was apparently developed in India, around the fifth century. From there it made its way relatively quickly east to Cambodia (and later to China), helping mathematize the Khmer—who were heavily influenced by Indian culture, including Hinduism, at the time—and others in novel ways.[11]

The journey west was a more languid one. The symbol for zero that we use as a place-holder (i.e., as a convenient marker of nothingness in mathematical operations) did not exist in Europe until the thirteenth century. A Persian mathematician named Muhammad Al-Kwarizmi (whose name is the source of the word *algorithm*) advocated the usage of the Hindu written numeral system, including zero, in his influential works written in the ninth century. Several centuries later those works were translated into European languages. In 1202, the Italian mathematician Leonardo of Pisa, more commonly known as Fibonacci, wrote his famous *Book of Calculation* (*Liber Abaci*). This manuscript also extolled the virtues of the usage of zero (and Hindu numerals more generally), suggesting that it facilitated a variety of mathematical procedures. Despite the reluctance of many Europeans to accept this Eastern system, eventually zero and decimal-based numerals did work their way into Western culture, becoming the dominant mathematical symbolic practice. Quite plausibly, the adoption of zero contributed to the subsequent refinement of European science and technology. Certainly a case can be made that this simple cognitive tool, an invented written numeral, yielded powerful effects on the lives of the Khmer, the Chinese, Europeans, and most humans today. After all, the facilitation of mathematical problem-solving means the facilitation of architecture, the facilitation of science, and in general represents a clear boon to technological development. While the zero-using decimal numerals Europe adopted may not be more "evolved" in a culturally neutral sense, they did seem to quicken the solution of certain kinds of mathematical tasks European cultures undertook during the latter portion of the Middle Ages and during the Renaissance. Europeans including the ancient Romans and Greeks had elaborate math without zero, and the symbol is hardly a necessary component of large civilizations. But it is also difficult to imagine the industrial or technological revolutions happening when they did without the use of the numeral.

Refinements of numerical language, including written refinements like zero, enable or at least accelerate language-external cultural changes. Consider how zero ameliorated Western mathematics. Without it, it would have been less convenient to symbolically depict negative numbers, the Cartesian plane, the graphing of functions, limits in differential calculus, and so on. In turn, these symbolic tools served as platforms for other mathematical strategies. It is probably not a coincidence that a mathematical revolution of sorts occurred in Europe after the introduction of zero notation. It also seems unlikely a coincidence that the introduction of zero and decimal-based mathematical writing neatly predates the earlier technological innovations of the Khmer empire. The faces of Bayon, then, represent a microcosm of the larger phenomenon we are highlighting: culture, in particular elaborate material culture, is influenced in extensive ways by invented numerical tools that are, in turn, the product of particular cultural traditions.[12]

Numbers at the Heart of Symbolic Innovation: Back to Writing

Writing was developed independently only a few times in human history—in Mesopotamia, in Mesoamerica, in China, and (debatably) in Egypt.[13] In each of these four traditions, the earliest known samples of writing are largely numeral-centric. This point was stressed in Chapter 2 with respect to the oldest writing system, that of Mesopotamia. Many of the earliest cuneiform tablets uncovered there are records of quantitative data. Full-fledged cuneiform seems to have only emerged after, or at least alongside, the development of systems of numeric bookkeeping.

Intriguingly, though, the same may be true of Chinese writing, the earliest samples of which date to the Shang Dynasty and are more than 3,000 years old. The most ancient of these samples are oracle bones. These bones were inscribed with numerals quantifying such items as enemy prisoners, birds and animals hunted, and sacrificed animals. In Mesoamerica, furthermore, a key shared commonality across the earliest scripts of the region is the representation of numerals with lines and dots. (See Figure 2.4.) The earliest written forms in the region are typically calendrical and numerical to one extent or another. Finally, in Egypt the oldest known hieroglyphic forms frequently convey information about the quantities of goods. It seems clear, then, that numerals are common to all the earliest forms of writing, not just that of Mesopotamia. Ancient writing around the world is numerically focused, much like ancient Paleolithic quasi-symbolic carvings and paintings are also often fixated on quantities. We saw this in Chapter 2 when we discussed artifacts like the 10,000-year-old antler found at Little Salt Spring. Such iconic forms are not as abstract or conventionalized as writing, but they clearly served a related function of conveying ideas in two dimensions.

So numerals have been present at the dawn of all writing systems. One reasonable interpretation of this fact is that written numerals serve as necessary precursors to more complete writing systems. But, if so, why do they play such a pivotal role in the emergence of writing? Let me offer one possible account, presaged in my discussion of ancient tally systems in Chapter 2. Consider a Roman numeral such as III. This numeral is iconic, with each line directly representing one element in a one-to-one arrangement. In contrast, the Latin word *et* ('and') represents a simple concept, but it is not iconic, because the two symbols in the word bear no actual physical relationship to what they represent, they do not match up with their referents like each symbol in III matches up with its referent. (If I say Super Bowl III, for example, each vertical mark represents one game.) Eventually numerals may become less iconic, so that the iconic origins of 7, for instance, are no longer evident. Numeric written symbols are likely easier to develop initially, when contrasted to symbols for other concepts and sounds, because of the notion of one-to-one correspondence. Singular quantities can be represented with singular lines, for example, and then larger quantities can be represented via the combination of these lines. More quantities require more lines (or dots or angles, etc.) in an iconic matching system. The inherent iconicity of many numerals, then, is based on our ability to recognize one-to-one correspondence. As has been stressed in previous chapters, this ability is innate for small quantities and is commonly acquired via language in the case of larger quantities. I have also stressed the role of fingers in the development of numerical thought, noting that fingers serve as the first linear representations of quantities in a person's life. Linear marks on stone, paper, or wood may be appreciated as representations of quantities more readily, because we are naturally exposed to linear representations of quantities on our hands. As a result of such factors, quantities can be represented directly and iconically via combinations of two-dimensional forms, and with a degree of cognitive ease that does not apparently hold for other concepts.[14]

So symbols for quantities can apparently be relatively easily written for at least three interrelated reasons. First, humans are innately predisposed to the abstraction of some quantity correspondences. We naturally recognize that items can match each other in simple yet abstract patterns of correspondence. Second, this abstract correspondence is relatively simple to signify with nonverbal symbols. After all, symbols for quantities do not require elaborate drawing or carving. In contrast, early representations of other concepts were written via pictograms in all written traditions, so that most symbols required greater intricacy than did numerals. 'Mammoth' or 'hunt,' for instance, were more difficult to symbolize than, say, I, II, or III. Third, our fingers serve as natural linear symbols that we first use for quantity correspondence. The usage of fingers as numerical icons likely facilitates the subsequent usage of other linear symbols for numbers. With time, these other iconic symbols can then become conventionalized and more abstract, so that true numerals gradually develop out of marks in tally systems.

In short, the inherent ease of representing quantities with lines and other marks may serve as a natural foundation for the more complete and abstract two-dimensional representation of quantities. This latter sort of representation may accelerate the realization that other concepts can also be represented two-dimensionally in abstract form. At the least, it is important to recognize that, in those few places in the world where full-fledged writing was truly invented (if gradually), written numerals were present at the genesis of the writing systems in question. Just as numbers were essential to the development and spread of agriculture, then, they seem to have been essential to the invention and spread of writing.[15]

Finally, the important role of numbers in the origins of writing is likely due to another simple fact too: numerals are incredibly practical. They serve essential functions in some kinds of human interactions, for instance, in economic transactions. Many of the first written records are the work of bookkeepers concerned with trades between two or more parties. Such bookkeeping facilitated the maintenance of trade networks and the careful storage of goods. Relatedly, numerals enabled calendrical practices allowing for the careful predictions of seasons and crop yields. Numerals are essential to many such activities of populous societies. (Societies whose origins can be traced, in turn, to agricultural practices facilitated by numbers.)

For such reasons, numbers were quite likely foundational to the advent of writing around the world. It is commonly recognized that the scientific revolution, industrialization, and modern medicine were dependent on specific mathematical practices. Millennia prior to the existence of these practices, though, verbal numbers and inscribed numerals helped enact profound changes in how humans subsisted and in how they used symbols to convey ideas.

Conclusion

From the pyramids of the ancient Egyptians, to the stone cities of Angkor, to the ruins of ancient Mesopotamia and Mesoamerica, a commonality surfaces. The agriculturalists who created these vast monuments relied heavily on numbers and, more specifically, on numerals. The earliest forms of their writing systems were focused to a surprising degree on the representation of numbers. Subsequently, numerals enabled new forms of engineering and architecture that changed the environments in which cultures developed. Numerals such as zero facilitated the manipulation of quantities. New kinds of cultural practices then developed, and these practices, in turn, placed newfound pressures on number systems. Prior to all that, elaborate verbal numbers were apparently crucial to the origin of certain kinds of agriculture—as evidenced, for example, by the fact that most contemporary hunter-gatherer populations employ number systems with low limits and restrictive functions. In short, spoken numbers and written numerals were pivotal to radical transformations in a variety of cultures millennia ago. In many contemporary endangered cultures, similar transformations are at work today.

10

TRANSFORMATIVE TOOLS

Once framed by my rearview mirror, Table Mountain has long since disappeared. Skirting the southernmost points of the African land-mass through meandering valleys, the road signs and radio adver-tisements are a collage of Afrikaans and English as I near my destination—a tranquil village named Stilbaai that hugs a stretch of azure waters just east of the divide between the Atlantic and Indian oceans. We do not know where numbers first originated, but it is possible that I am approaching the place. It may be that the story of numbers began here, along a stretch of craggy coast-line visible from Stilbaai—coastline that is increasingly being recognized as a key setting in the human narrative.

For the past two decades, archaeologists of various specialties have been combing the floors, and the former floors now under-ground, of some caves near this coastline—Blombos Cave just west of Stilbaai and the Pinnacle Point caves dozens of kilometers to the east. The research based in these locales has produced eye-catching results. That research is, along with other studies on early *Homo sapiens,* shedding new light on how our ancestors survived and even excelled in the millennia preceding *sapiens'* African exodus. What has been found in these caves is not the sort of find one typically associates with African paleoarchaeology. There are no *australo-*

pithecine bones, nor are their skeletal remnants of any other potential ancestors of *Homo sapiens*. (Those have been discovered elsewhere in South Africa and in far-off African places like the Olduvai Gorge and the Great Rift Valley.) No major hominin remains have been found at Blombos and Pinnacle Point—only a few ancient human teeth and bone fragments. Yet what has been found probably tells us more than any skeleton could about the lives of our more immediate ancestors, members of our own species who looked and behaved much like us. Judging from the finds in these caves, much of what we now consider modern human behavior may have originated along this coastline. This includes some behavior consistent with the usage of numerical technologies.

From about 190,000 to 135,000 years ago, the global climate shifted. As with older climatic shifts, this change appears to have dramatically tilted the human stage. Previous climatic changes, such as the one which occurred about 1.9 million years ago, may have been pivotal to the genesis of our species by forcing our ancestral species to, for instance, live in savannahs instead of forests. The more recent shift in question impacted *Homo sapiens* itself, though, as it resulted in a pronounced reduction in the habitable regions of Africa. The continent became more arid, and our food sources scarcer. Additionally, about 75,000 years ago a massive super-eruption of the Toba volcano in Sumatra led to an enormous ash cloud and volcanic winter that may have reduced the population of humans significantly. The paleoarchaeological evidence suggests that humans took refuge in coastal regions during these difficult times, especially the southernmost coast of Africa. The likely motivations for this choice of refuge, recent findings demonstrate, is that this coast was comparatively rich in food sources. More specifically, the southern tip of Africa offered readily available seafood, such as marine snails, as well as geophytes—fleshy plants like bulbs and tubers that survive underground. While food sources were relatively scarce in many human homelands at the time, here the land was rich in both carbohydrates and proteins. The climatic and ecological reconstructions of the era in question are fairly unequivocal with respect to this point—this coastline was a good place for people to live during an era when there were apparently few habitable places in Africa.[1]

Judging from the finds at Pinnacle Point, people did not just survive here, they thrived in this region during the period in question. Humanity witnessed a kind of technological florescence while people lived along this coastline, beginning about 170,000 years ago. Archaeologists have now shown that, over the tens of thousands of years the Pinnacle Point caves were used, humans refined technologies, such as their stone tools. In those caves there is evidence of tools that were made after the source stones were heated in fire and then flaked, in a complex tool-fashioning process. In contrast to the stagnancy of stone-tool technology during preceding epochs, the finds at Pinnacle Point suggest a comparatively radical rate of innovation in the usage of new types of stone implements. Furthermore, at Pinnacle Point there are other traces of technological advances, such as red pigment that was possibly used for body painting as it is in some populations today. This sort of material residue demonstrates that technology was advancing during this period but also that people were thinking in symbolic ways and transferring material technologies across generations. Such residue suggests, therefore, that the ancient residents of this region possessed language.[2]

Spearheaded by archaeologist Christopher Henshilwood of the University of Bergen in Norway, a series of studies have shown that Blombos Cave was also used by humans over the course of millennia—probably for a period of about 30,000 years. The detritus at that cave is generally more recent and offers even more evocative evidence for the advancement of human cognition. This evidence includes refined stone tools and needle-like bone tools.

More tellingly, perhaps, it also includes bowl-like abalone shells, grindstones, and other items once used as ochre-processing kits for the extraction of pigment from the iron-infused mineral. In fact, Blombos Cave seems to have served as a workshop for the fabrication and processing of various sorts of artifacts. These artifacts include engraved pieces of bone and ochre dating back 100,000 to 70,000 years before the present. On the most famous find taken from the cave, a piece of ochre only about 6 centimeters long, archaeologists observed a series of regular hatchlike marks that were clearly carved intentionally by a human artist. It is unclear what the purpose of these marks was, but they may have served some symbolic or quasi-symbolic function. This is perhaps the oldest human artifact that is symbolic in nature. One possibility, given the extensive evidence for prehistoric numerals throughout the world's Paleolithic settings, is that the marks served some quantitative function. Could it be that the artisan who engraved the ochre was recording some quantity, like the engravers of the Ishango bone tens of millennia later? (See Chapter 2.) Unfortunately, the true function of the piece is probably lost to the turbid archaeological record.[3]

Other evidence in Blombos Cave also hints that the people who used it as a workshop may have developed ways of recording quantities—they may have invented or at least inherited numbers. Among the remarkable finds from the locale are many perforated marine shells (so-called tick shells of the species *Nassarius krausianus*), found in small sets like five and twelve shells. The shells, typically about a centimeter long, were apparently used as personal adornments. The neatly aligned human-made perforations allowed people to tie the shells together into necklaces or some other kind of ornament, just as in many cultures today. Tellingly, some of the shells were not native to the immediate area in which the cave is located, even all those millennia ago. Today these particular shells are only found in estuaries about 20 kilometers from Blombos.[4] So the people who lived in this region appear to have valued these lustrous shells immensely, to the point that they traveled a long distance by foot to find them, or potentially traded with other people for them. What we have evidence for at Blombos, then, is the ancient usage of small, relatively homogeneous artifacts of great value.

It is possible that the people who lived near the cave would have faced strong pressures to invent ways of recording quantities symbolically, pressures to invent numbers, perhaps to keep better track of these valuable shells, to trade quantities of other commodities in exchange for them, or both. Some researchers have gone so far as to suggest that these shell beads actually served as symbolic representations of quantities (i.e., that they were numbers themselves). As the findings discussed in this book hopefully make clear, however, it is more probable that fingers rather than beads served as the first representations of precise quantities—as the first numbers. As prominent psychologist Susan Carey (whose influential work on numerical cognitive development is discussed in Chapter 6) notes in discussing the importance of these artifacts: "beads may go back 100,000 years, but fingers go back millions."[5] Fingers typically serve as the gateway to the recognition of precise quantities and are often used as numbers that are then instantiated verbally. Still, it seems likely that small, similar, discrete items of great value at some point led humans to *want* to quantify things in nonapproximate ways, even if they needed their fingers to do so. Perhaps the pressures to quantify these lustrous beads in Blombos Cave created a new need for numbers, a newfound desire to consistently and exactly discriminate quantities.

While we cannot definitively establish the first place where numbers were used, the picture being painted here seems reasonable. The people who lived in this coastal region possessed material culture, language, and maybe even two-dimensional symbolic practice. They also had valuable, miniscule commodities that they probably wanted to count, given the lengths to which they went to procure them. In the light of these facts, it is not implausible that they did have numbers. But, assuming for the moment that was indeed the case, were numbers invented here or brought here? Given the pace at which technology began evolving along this coastline, judging from the finds at Pinnacle Point and Blombos, the former scenario is at least possible. Maybe humans refined their linguistic and numeric skills here, on this stretch of coastline and in other parts of southern Africa. Such refinement, in turn, may have played a pivotal role in the ability of humans to adapt to various other environs. Certainly the usage of language, unambiguously evident in the buried artifacts at Blombos and Pinnacle Point, enabled us to subsequently conquer Africa and leave the continent en masse.

Arriving at the shoreline that apparently played such an important role in human history, I find that it is speckled with countless boulders. (See Figure 10.1.) These boulders have not been coastal at all times, as the shoreline itself shifts somewhat with the comings and goings of ice ages. Yet they were about as near to Blombos Cave 100,000 years ago as they are now. Many of the boulders are nearly cubelike in shape, almost as if human-made. But, if so, they were arranged by some inebriated architect and placed slightly askew of one another. Stepping among them, one is reminded of the locals who climbed over and around these same stones all those dozens of millennia ago. Here on this shore people saved themselves through a newfound taste for seafood. More broadly, the decisions they made on these rocks, in nearby caves, and elsewhere in this region may well have saved our species from perishing. Certainly those decisions seem to have facilitated our continued survival during hard times and our subsequent expansion out of Africa.

Scattered around the boulders is circumstantial evidence for the pressures these ancestors of ours would have faced to develop

10.1.   The coast near Blombos Cave, South Africa. Photograph by the author.

numbers. It is weak evidence, admittedly, but an indirect hint lying around the rocks periodically swallowed up by the inexorable surf: striated lavender and white shells, dispersed with some regularity. We now know that such shells and the gastropods they contained, still as obtainable as they were all those millennia ago, were vital to the survival of humans during those lean times. They were also a crucial thread in the fabric of the local material culture. Perhaps the inherent value of such countable items empowered someone to enumerate the shells. Maybe she recognized that the shells could be aligned just like the fingers of her hands were symmetrically aligned, or realized straightaway that five shells could correspond in a one-to-one manner to the fingers of a hand. Maybe that recognition was brought to life verbally as she began speaking of a 'hand' of shells. Perhaps the recognition was made with greater ease thereafter, by herself and by other members of that population who became practiced in the concept of a 'hand' of shells or a 'hand' of other items. We do not know, of course, but at this point this seems a more likely setting than any to have been the place where people first began using numbers.

What we do know is that someone, somewhere, at a particular moment in history, was the first person to abstractly recognize the concept of exactly five. Yet this recognition, crucial to the invention of number systems, no doubt occurred many other times independently and in various cultural lineages. It was likely lost to time in some cases. But in other cases, the fickle recognition was reified symbolically, made real with a word. This word was then transferred to other minds that built on the concept in new ways. While the first inventors of number words like 'five' had no appreciation of this fact, their newly fashioned cognitive tools would one day shift the course of human cultures.

Numbers and Deities

Our obsession with the number of our days and years, which is evident in larger societies from the ancient Maya to modern Americans, partially stems from agricultural practices that in turn owe their existence to the creation of number systems. A shift to agricultural subsistence was also the forerunner of other sorts of more intimate changes to the human experience—changes not just to how we tally our age, but to how we see our place in the universe. Here I am not referring to the fact that numbers and agriculture yielded a greater reliance on the tracking of the stars, seasons, and so on, and therefore eventually led to astronomical awareness and an appreciation of the non-human-centered universe in which we live. While the latter is no doubt true, I am referring instead to the new kind of religious enterprises that followed the elaboration of number systems.

This may seem a bit of a stretch, ascribing a spiritual significance to numbers. To be sure, all populations seem to have some form of religious belief and spirituality, regardless of the kinds of number systems they employ. But the point here is a more nuanced one, and one that is supported by the archaeological and anthropological record: although creation myths, animistic practices, and other forms of spiritualism are universal or nearly universal, large-scale hierarchical religions are restricted to relatively few cultural lineages. Furthermore, these religions—including monotheistic ones, such as Islam, Christianity, and Judaism, but also other major world religions, such as Hinduism, Shintoism, and Buddhism—developed long after agriculture. More to the point, they developed only after people began living in larger groups and settlements because of their agricultural lifestyles. For the bulk of our 100,000-plus year existence, our species lived in small bands or tribes in places like Blombos. In the past 10,000 years or so, and in particular the past several millennia, we have congregated into larger chiefdoms and empires, often with major urban areas at their cores. A phalanx of scholars has recently suggested that the development of major hierarchical religions, like the development of hierarchical governments, resulted from the agglomeration of people in such places. Assuming for the moment that their reasonable hypothesis is correct, this suggests that the innovation of complex number systems, by facilitating agriculture, ultimately resulted in new perspectives on the role of humans in the world—new views on the origins of the earth and those on it. We could go so far as to say that the development of number systems was pivotal to the creation of God(s). Or perhaps, to some, this development led to the accurate realization that God(s) exists.

This conjecture is based on the claim that larger populations did causally yield new, typically theistic, religious traditions. Why, then, might have larger populations led to theism? In broad strokes, the hypothesis in question goes something like this. Organized religious beliefs, with moral-enforcing deities and priest classes, were a by-product of the need for large groups of people to cooperate via shared morals and altruism. As the populations of cultures grew after the advent of agricultural centers and associated urbanization, individuals were forced to rely on shared trust with many more individuals, including non-kin, than was or is the case in smaller groups like bands and tribes. This shared trust within cultures was essential if those groups were to outcompete other similarly sized groups of individuals. In contrast, groups like bands and tribes were (and are) small, and most individuals in a band of hunter-gatherers have some kinship-based relationship to all or most of the other members of their culture. So the natural motivations for trust and cooperation among individuals are clearer in the case of small populations. Since natural selection is predicated on the protection of one's genes, in-group altruism and sacrifice are easier to make sense of in bands and tribes. But why would humans in much larger populations—humans who have no discernible genetic relationship to the majority of individuals they come into contact with on a daily basis—cooperate with these other individuals in their own culture? Why would they concern themselves with the well-being of complete strangers through continually cooperative acts? According to the hypothesis in question, championed by researchers including psychologists Ara Norenzayan and Azim Shariff, some social mechanism had to evolve so that larger cultures would not disintegrate due to competition among individuals and so that many people would not freeload off the work of others. One social mechanism that fosters prosocial and cooperative behavior is an organized religion based on shared morals and omniscient deities capable of keeping track of the violation of such morals. The gradual development of such deity-centered religions may have served to naturally maintain order and cooperative behaviors that benefited cultural survival and success through the prosocial concern for others. Framed differently, large human populations without deity-centered moralistic religions would have been less likely to survive when confronted with other large groups that, while not particularly cooperative with outsiders (and perhaps possessing a certain degree of bloodlust toward outsiders), were more internally cooperative due to a theistic religion that enforced cooperation with non-kin in their own population.

Some support for this view comes from a recent survey of many of the world's cultures. According to that survey of 186 contemporary societies, there is a strong correlation between the population size of a given culture and the likelihood that the culture maintains a religion centered around a deity (or deities) who concerns itself with the morality of individuals. Such correlational work is not conclusive, of course, but does hint that this account is on the right track. What is clear is that the growth of large hierarchical religions based around God(s) is a fairly recent trend. Furthermore, it is a trend that followed the numerically influenced agricultural revolution that enabled the growth of human populations in the regions where the religions in question then developed. These "new" religions in turn have transformed the views of many humans regarding their place in the universe, altering their worldviews, and imbuing them with a distinct sense of purpose. Because of this turn of events, many people have come to see themselves as the special creations of God(s). At least indirectly, the development of elaborate number systems played a role in the transformation of humans' understanding of their own souls.[6]

Socially Significant Numbers

When Moses descended from Mt. Sinai with his stone tablets, they were inscribed with ten divine moral imperatives. Ten commandments. Even if you do not subscribe to one of the religious traditions that acknowledges these commandments, you are aware of their existence. While you may not be able to recite all ten, you do know there are ten of them. Why ten? Surely there are more than ten religious imperatives that could have been doled out to humanity—more specifically, to a group of middle-eastern nomads—more than 2 millennia ago. Here is an eleventh commandment that could likely be uncontroversially adopted by many people: "thou shalt not torture." One suspects that, had Moses come down with this as an inscribed commandment, there would have been no uproar. Likewise, today most people would agree with the sentiment of this missing commandment. But then the list would appear to lose some of its rhetorical heft. "Eleven commandments" almost hints of a satirical deity. Had Moses come down with eleven commandments, his audience may not have objected to the eleventh commandment per se, but the existence of precisely eleven may have struck some of them as odd. Children learning the eleven commandments in religious contexts might be more confused than if they were learning the holier number of ten. Ten is the roundest of numbers, and it makes sense to many that the rules governing our lives would come in a package of ten. Interestingly, the number ten is also of spiritual import across a variety of other religious traditions: there are the ten avatars of Vishnu, the ten human gurus in Sikhism, the ten attributes in Kabbalah, and so on. This recurring influence of ten is not, it seems, coincidental. Maybe we should not be surprised to find that ten is commonly ascribed special spiritual and social import, given its widespread influence in number systems. The ultimate motivation for this import is by now clear: it is not so much that divine concepts come in packages of ten, but that our fingers do. (To some, of course, this finger-packaging itself may be divine.)[7]

Other significant numbers in religious texts are also often nice and round, neatly divisible by ten. For instance, 'forty' plays an important role in the Judeo-Christian tradition: Noah's flood lasted 40 days, Jesus wandered in the desert for 40 days, Moses spent 40 days on Mt. Sinai, Elijah fasted for 40 days, Jesus ascended 40 days after his crucifixion, and so forth.[8]

Alongside the widespread finger bias evident in the significant and holy numbers of some religious traditions, an interrelated phenomenon at work here is the injection of numbers with spirituality. This is true in the world's major religions with holy numbers and in other common spiritual traditions besides those mentioned so far. Certain Chinese traditions consider a variety of numbers to be auspicious or inauspicious. Proponents of astrology, meanwhile, maintain that particular numbers reflect specific spiritual or personality traits or both.

A neat circularity is evident in these sorts of conventions. As noted above, agricultural and sedentary lifestyles yielded larger human populations, which in turn are causally implicated in the origin of many theistic and moralistic belief systems. Given that number systems helped produce agricultural practice, they are at least partially responsible for religious and spiritual practices that have, in turn, infused specific numbers with religious and spiritual significance. This pattern of cultural feedback is characteristic of the general story of coevolution suggested at various points in this book—numbers and counting fundamentally impacted the human experience, and the results of their impact then placed new pressures on how, and to what extent, people rely on numbers. These included pressures to spiritualize numbers.

Ascribing spiritual and social significance to particular numbers may strike some as a quaint vestige of pre-scientific times. Yet, arguably now more than ever, numbers are ascribed significance as they have become associated, unreflectively in some instances, with scientific discoveries. After all, mathematics has played a sizable role in shaping the dramatic scientific progress of the last few centuries. Most people recognize that modern science is based on results that are usually quantified. The scientific method, so tied to math of various sorts, is understandably perceived as the starting point on a path to higher truth. In that sense, atheistic or agnostic proponents of science may also imbue numbers with a kind of spiritual importance, treating them as mind- and body-external realities that guide us toward discoveries of new truths. While the quantities numbers represent may exist outside our minds, however, the symbolic representations of those quantities are our own innovations, not truly divorceable from our minds. And scientific practices rely on a kind of spiritualization of those anatomically contingent innovations. In many modern societies numbers are epistemologically crucial—they help tell us whether something is a justifiable belief. And the numeric description of new beliefs injects them with special kinds of meaning. For instance, if cosmologists inform us that the universe is really old, that may mean something. But if they tell us the universe is more than 13 billion years old, that may mean much more to us—regardless of whether we can truly conceptualize that sort of time scale. Once described with numbers, observations seem more palatable as facts.

Yet numbers are not just ascribed social value in this general epistemological sense. Some specific numbers still brandish social and spiritual-like significance in nonreligious contexts—significance that is more arbitrary than many observers realize. This is perhaps most plainly evident in the usage of so-called significant values. These values are fascinating, because they illustrate that, in supposedly asocial and aspiritual contexts, certain numbers have socially sanctioned value that gives them, often without reflection, a special place in our minds. And that special place is ultimately due, once again, to the relationship of numbers to our anatomy.

Here is what I mean. If you pick up any science journal you will likely find that many or most articles have what are called P values scattered throughout. Researchers become familiar with these values early on in their training. (So excuse the superfluity here, if you are a researcher of some kind.) These values are derived from various kinds of statistical analyses based on the results of experiments or other forms of data collection. These P values reflect the odds that a specific result is due to the null hypothesis, rather than the hypothesis being tested. So, for instance, a study might examine the correlation between smoking rates and lung cancer prevalence in a given population. After testing the strength of the correlation, the researchers might conclude that P is less than, say, 0.004. This would imply that the null hypothesis (in this case that smoking and lung cancer have no relationship) is very unlikely. That is, the odds of the null hypothesis holding are less than 4 in 1,000 in this made-up case. Low P values suggest that the results obtained in a given study are probably not due to chance, and offer support for the tested hypothesis. In the decades since the 1920s, when statistician Ronald Fisher introduced P values, they have come to play a ubiquitous role in science. Some readers of academic publications skim articles initially, looking for the P values and trying to get a quick grasp of the strength of the results being discussed. People want to know right away if the findings are "significant." Researchers often hope for low, significant P values when they are conducting analyses. Low P values usually increase the likelihood of publication of their work, of future funding, and so on. Setting aside the ways in which P values have arguably been overapplied and misapplied since Fisher's work on the topic, as well as the disagreements that statisticians have over their usefulness, it is indisputable that P values matter a great deal in the contemporary scientific community. They give the results of many studies meaning, or at least allow people to take meaning away from them more readily.[9]

Yet in some ways that meaning is chimerical, or at least not as profound as some readers of scientific works may perceive it to be. To get a sense why, consider what counts as "significant" when considering P values. If P is less than 0.01, the results of a study are generally considered to be very significant. At some point during the past few decades, P values of less than 0.05 came to be considered statistically significant as well in some fields, though less significant than values below 0.01. A P value less than 0.05 implies that there is less than a 5 in 100 chance that the null hypothesis of a given study is true. But why 5 out of 100? Or 1 out of 100? Did the universe impartially establish such ratios as a pathway to knowledge, to truth? Of course not. Instead, what we see in such P values is a by-now-familiar pattern. Five and ten, and multiples of these numbers, are special to us. We are prone to ascribing social significance to these numbers not because they are more objectively tied to scientific truth, but because they are tied to our hands. Even if we do not think about the manual basis of this quinary and decimal bias, it is always there in a historical sense. The truth is that much of science, or more precisely, the way in which scientific studies are interpreted by many, is unreflectively manually based. P values could be socially sanctioned at other ratios. Maybe P should be less than 0.03 to be considered significant. Or 0.007. Or 0.023. These are low and arbitrary figures, but as justifiable as 0.05 or 0.01 from an impartial non-finger-biased perspective.

All we can really establish is that, when the P values arrived at in given study are low, the odds of the null hypothesis being true are not good. But this gives us less assurance. We want the numbers to tell us, simply, what the meaning of a study is. This is not to suggest that the statistical tests are not elucidative, but to point out that the way we interpret them is usually not as subtle as it could be. Instead, how we interpret scientific studies is often an all-or-nothing game of "significant or not significant," and this dichotomous choice does have its benefits in simplifying our reading of the data. Crucially, though, the way we make the choice in question is fundamentally just the by-product of the characteristics of the hands we do the science with. In a way, P values are the latest example of how we, often without realizing it, use our fingers to point to higher truth—not too unlike the ten commandments.

Conclusion

Potentially as far back as our ancestors' seaside residence near present-day Stilbaai, numbers have been refashioning the human experience. They continue to do so, from temples and churches to universities and laboratories. They still alter the lives of large agricultural populations, as well as those of small hunter-gatherer and horticulturalist groups now being forcibly integrated into an increasingly globalized existence.

Verbal and nonverbal representations of precise quantities have changed nearly every conceivable facet of our lives. As you read these words, little of your world—from your interior thoughts to your exterior environment—has not been affected directly or indirectly by these representations, by numbers. The neat lines and letters on this page would not be possible without numbers. After all, numbers enable measurement, and numerals were the precursor to writing. It would probably be less onerous to detail the ways our lives have not been changed by the innovation of number systems than the ways they have been. Everything from modern medicine, religion, and industrialization to architecture and athletics has been influenced by the invention and elaboration of these systems in ways that are often unrecognized.

In this book I have made the claim that numbers, the symbolic concretizations of quantities, are in fact an invention. Quantities exist in nature, even regularly occurring quantities—whether the years between the reproduction cycles of cicadas, or sets of arachnid legs, or days in a lunar cycle, and so forth. But numbers themselves, the symbolic incarnations of such regular quantities, do not exist apart from human creation. And we do not create them, simply, with innate mechanisms. This claim is based on recent evidence from infants, from anumeric people, and from species related to our own. As we have seen, all of that evidence converges on a clear conclusion: we are not born with the ability to distinguish most quantities precisely, though we have innate capacities that allow us to approximate quantities and to exactly distinguish small sets of items. These capacities do not allow us to numerically differentiate most sets of things or events, though, even those sets found in nature. It was the innovation of numbers, representations of specific quantities, that enabled people to consistently appreciate quantitative patterns in precise ways. Prior to the invention of number systems, most of the quantitative regularities of nature were hidden from the view of *Homo sapiens,* or any other animal for that matter. Their invention resulted in a seismic cognitive shift, one whose aftershocks are ongoing.

I have also suggested that the invention of most types of numbers was not just a natural by-product of language and culture alone, but was due to the biological symmetry of human hands, hands that we can readily fixate on and that are not required for locomotion. Our bipedalism enabled a greater focus on, and refined manipulation of, our hands. This focus and manipulation eventually resulted in the occasional recognition of the quantitative correspondence of the fingers to each other and to other items. This simple recognition, something so simple yet not prewired in us, was eventually brought to life linguistically. Numbers came into being. Precise quantities became a constant presence in our thoughts rather than being encountered only sporadically. Judging from the archaeological and linguistic records, precise quantities have been conveyed consistently through numbers for many millennia. In that sense, the numerical revolution is an old one.

Yet, in another real sense, the numerical revolution only gained steam in the past few thousands of years. During that time, number systems and agriculture coevolved. The results of this coevolution include larger societies, particular kinds of religious beliefs, mathematics, and writing systems that were initially number-centric in all cases. Inarguably, numbers and counting changed the human story. While people have long recognized the importance of the development of mathematics in human history, I have stressed that the invention of number words, spoken numbers, played a more substantial and earlier role. In line with much recent cross-disciplinary research on this topic, I have suggested that such numbers were and remain cognitive tools—tools that transformed our lives long before the usage of advanced mathematics.

As human beings we are constantly floating in a sea of quantities, much as we bob in a sea of other stimuli. We float in a sea of visible light, for example. Just as eyes allow us to differentiate that light and navigate the physical world around us, numbers help us differentiate the quantities around us and traverse new conceptual waters. As I have stressed in this book, these tools for conceptual navigation do not exist apart from our creation of them. We did not discover them lying around somewhere on the coast of southern Africa. Instead, at various times and places, including perhaps the coast near Stilbaai, we took realizations of quantity correspondences and reified them by creating new kinds of words. Typically, the correspondences were between quantities in the wild and quantities representable with our fingers.

In essence, then, we took our hands and reached into the sea of indistinguishable quantities, shaping them into numbers. We took hold of quantities of items around us—both metaphorically and literally. We molded abstract quantity correspondences into very real, but very unnatural, numbers. We made numbers. And given their transformative effect on the human story, it is fair to say that numbers also made us.

NOTES

ACKNOWLEDGMENTS

INDEX

NOTES

Prologue

1. For some accounts of shipwrecked sailors surviving with indigenous cultures, see Alvar Núñez Cabeza de Vaca, *The Shipwrecked Men* (London: Penguin Books, 2007).

2. See, for example, Brian Cotterrell and Johan Kamminga, *Mechanics of Pre-Industrial Technology* (Cambridge: Cambridge University Press, 1990).

3. For more on the cultural ratchet, see Claudio Tennie, Josep Call, and Michael Tomasello, "Ratcheting Up the Ratchet: On the Evolution of Cumulative Culture," *Philosophical Transactions of the Royal Society B* 364 (2009): 2405–2415, as well as Michael Tomasello, *The Cultural Origins of Human Cognition* (Cambridge, MA: Harvard University Press, 2009).

4. For discussion of this Inuit case, and for elaboration of the notion of culturally stored knowledge, see Robert Boyd, Peter Richerson, and Joseph Henrich, "The Cultural Niche: Why Social Learning Is Essential for Human Adaptation," *Proceedings of the National Academy of Sciences USA* 108 (2011): 10918–10925. For more on the evolution of cultures, see, for example, Peter Richerson and Morten Christiansen, eds., *Cultural Evolution: Society, Technology, Language, and Religion*. Strüngmann Forum Reports, volume 12 (Cambridge, MA: MIT Press, 2013).

1. Numbers Woven into Our Present

1. For more on the perception of time among the Aymara, see Rafael Núñez and Eve Sweetser, "With the Future behind Them: Convergent Evidence

from Aymara Language and Gesture in the Crosslinguistic Comparison of Spatial Construals of Time," *Cognitive Science* 30 (2006): 401–450.

2. Thaayorre temporal perception is analyzed in Lera Boroditsky and Alice Gaby, "Remembrances of Times East: Absolute Spatial Representations of Time in an Australian Aboriginal Community," *Psychological Science* 21 (2010): 1621–1639.

3. In a related vein, it is worth noting that the duration of the earth's rotation (whether sidereal or with respect to the sun) is not absolute. For instance, prior to the moon-creating collision of a planetesimal with the earth billions of years ago, the earth's solar day lasted only about six hours. Even now days are gradually increasing in duration as the rotation of the earth slows bit by bit due to tidal friction, and furthermore solar days vary slightly depending on the earth's orbital position relative to the sun. For more on this topic, see, for instance, Jo Ellen Barnett, *Time's Pendulum: From Sundials to Atomic Clocks, the Fascinating History of Timekeeping and How Our Discoveries Changed the World* (San Diego: Harcourt Brace, 1999).

4. It is also the result of the development of associated mechanisms used to keep track of time, from sundials to smart phones. Interestingly, this development reflects the increasingly abstract nature of time-keeping. Where once such mechanisms, like sundials and water clocks, were used to track the diurnal cycle, they eventually came to track units of time that are independent of celestial patterns. This transition stems in part from the development of weight-based clocks (particularly pendulum clocks) and spring-based time pieces, which allowed for more accurate measurement of time than any celestial methods available. Such accurate time measurement enabled, among other major innovations, more precise longitude measurement and navigation. See the fascinating discussion in Barnett, *Time's Pendulum*.

5. There are many excellent books on human evolution and paleoarchaeology. For one recent exemplar, see Martin Meredith, *Born in Africa: The Quest for the Origins of Human Life* (New York: Public Affairs, 2012).

6. The claims regarding australopithecines are based on the famous work of the Leakeys, notably in Mary Leakey and John Harris, *Laetoli: A Pliocene Site in Northern Tanzania* (New York: Oxford University Press, 1979), as well as Mary Leakey and Richard Hay, "Pliocene Footprints in the Laetolil Beds at Laetoli, Northern Tanzania," *Nature* 278 (1979): 317–323. See also Meredith, *Born in Africa*.

7. Some of the research in the Blombos and Sibudu caves is described in Christopher Henshilwood, Francesco d'Errico, and Ian Watts, "Engraved Ochres from the Middle Stone Age Levels at Blombos Cave, South Africa," *Journal of Human Evolution* 57 (2009): 27–47, as well as Lucinda Backwell, Francesco d'Errico, and Lyn Wadley, "Middle Stone Age Bone Tools from the Howiesons Poort Layers, Sibudu Cave, South Africa," *Journal of Archaeological Science* 35 (2008): 1566–1580. The location of the African exodus is taken from the synthesis in Meredith, *Born in Africa*.

8. The antiquity of humans in South America, more specifically, Monte Verde in present-day Chile, is discussed in David Meltzer, Donald Grayson, Gerardo Ardila, Alex Barker, Dena Dincauze, C. Vance Haynes, Francisco Mena, Lautaro Nunez, and Dennis Stanford, "On the Pleistocene Antiquity of Monte Verde, Southern Chile," *American Antiquity* 62 (1997): 659–663.

9. The cooperative foundation of language is underscored in, for example, Michael Tomasello and Esther Herrmann, "Ape and Human Cognition: What's the Difference?" *Current Directions in Psychological Science* 19 (2010): 3–8, and Michael Tomasello and Amrisha Vaish, "Origins of Human Cooperation and Morality," *Annual Review of Psychology* 64 (2013): 231–255.

10. For more on how language impacts thought, see, for example, Caleb Everett, *Linguistic Relativity: Evidence across Languages and Cognitive Domains* (Berlin: De Gruyter Mouton, 2013) or Gary Lupyan and Benjamin Bergen, "How Language Programs the Mind," *Topics in Cognitive Science* 8 (2016): 408–424.

11. For a global survey of world color terms, see Paul Kay, Brent Berlin, Luisa Maffi, William Merrifield, and Richard Cook, *World Color Survey* (Chicago: University of Chicago Press, 2011). The experimental research conducted among the Berinmo is reported in Jules Davidoff, Ian Davies, and Debi Roberson, "Is Color Categorisation Universal? New Evidence from a Stone-Age Culture. Colour Categories in a Stone-Age Tribe," *Nature* 398 (1999): 203–204.

12. Other terminological choices can be made here. One could refer to regular quantities as 'numbers,' rather than restricting the usage of the latter term to words and other symbols for quantities. If that terminological choice were adopted, however, the central point would be unaltered: Our recognition of precise quantities is largely dependent on number words.

13. Heike Wiese, *Numbers, Language, and the Human Mind* (Cambridge: Cambridge University Press, 2003), 762.

2. Numbers Carved into Our Past

1. The paintings at Monte Alegre are discussed in, for example, Anna Roosevelt, Marconales Lima da Costa, Christiane Machado, Mostafa Michab, Norbert Mercier, Hélène Valladas, James Feathers, William Barnett, Maura da Silveira, Andrew Henderson, Jane Silva, Barry Chernoff, David Reese, J. Alan Holman, Nicholas Toth, and Kathy Schick, "Paleoindian Cave Dwellers in the Amazon: The Peopling of the Americas," *Science* 33 (1996): 373–384. For a discussion of the possible calendrical functions of the particular painting mentioned here, see Christopher Davis, "Hitching Post of the Sky: Did Paleoindians Paint an Ancient Calendar on Stone along the Amazon River?" *Proceedings of the Fine International Conference on Gigapixel Imaging for Science* 1 (2010): 1–18. As Davis notes, famous nineteenth-century naturalist Alfred Wallace mentioned and sketched some of these Monte Alegre paintings in his work.

2. The antler was first described in John Gifford and Steven Koski, "An Incised Antler Artifact from Little Salt Spring," *Florida Anthropologist* 64 (2011): 47–52. The authors of that study note the possibility that the antler served a calendrical purpose, though some of the points made here are based on my own interpretation.

3. Karenleigh Overmann, "Material Scaffolds in Numbers and Time," *Cambridge Archaeological Journal* 23 (2013): 19–39. For one comprehensive interpretation of the Taï plaque, see Alexander Marshack, "The Taï Plaque and Calendrical Notation in the Upper Paleolithic," *Cambridge Archaeological Journal* 1 (1991): 25–61.

4. For one analysis of the Ishango bone, see Vladimir Pletser and Dirk Huylebrouck, "The Ishango Artefact: The Missing Base 12 Link," *Forma* 14 (1999): 339–346.

5. The Lebombo bone is discussed in Francesco d'Errico, Lucinda Backwell, Paola Villa, Ilaria Degano, Jeannette Lucejko, Marion Bamford, Thomas Higham, Maria Colombini, and Peter Beaumont, "Early Evidence of San Material Culture Represented by Organic Artifacts from Border Cave, South Africa," *Proceedings of the National Academy of Sciences USA* 109 (2012): 13214–13219.

6. For more on the world's tally systems, see Karl Menninger, *Number Words and Number Symbols* (Cambridge, MA: MIT Press, 1969). For a more detailed description of the Jarawara tally system, see Caleb Everett, "A Closer

Look at a Supposedly Anumeric Language," *International Journal of American Linguistics* 78 (2012): 575–590.

7. For detailed analysis of these geoglyphs, see Martti Parssinen, Denise Schaan, and Alceu Ranzi, "Pre-Columbian Geometric Earthworks in the Upper Purus: A Complex Society in Western Amazonia," *Antiquity* 83 (2009): 1084–1095.

8. Karenleigh Overmann, "Finger-Counting in the Upper Paleolithic," *Rock Art Research* 31 (2014): 63–80.

9. The Indonesian cave paintings, possibly the oldest uncovered to date, are discussed in Maxime Aubert, Adam Brumm, Muhammad Ramli, Thomas Sutikna, Wahyu Saptomo, Budianto Hakim, Michael Morwood, G. van den Bergh, Leslie Kinsley, and Anthony Dosseto, "Pleistocene Cave Art from Sulawesi, Indonesia," *Nature* 514 (2014): 223–227. For an example of how such cave paintings are dated, see the discussion of the Fern Cave in Rosemary Goodall, Bruno David, Peter Kershaw, and Peter Fredericks, "Prehistoric Hand Stencils at Fern Cave, North Queensland (Australia): Environmental and Chronological Implications of Rama Spectroscopy and FT-IR Imaging Results," *Journal of Archaeological Science* 36 (2009): 2617–2624.

10. Many books have been written on the history of writing. My claims here are based in part on Barry Powell, *Writing: Theory and History of the Technology of Civilization* (West Sussex: Wiley-Blackwell, 2012).

11. I am grateful to an anonymous reviewer for pointing out this example.

12. For more on this Sumerian history, and the history of other numeral and counting systems, see Graham Flegg, *Numbers through the Ages* (London: Macmillan, 1989) and Graham Flegg, *Numbers: Their History and Meaning* (New York: Schocken Books, 1983).

13. For a cognitively oriented survey of the world's numeral systems, see Stephen Chrisomalis, "A Cognitive Typology for Numerical Notation," *Cambridge Archaeological Journal* 14 (2004): 37–52.

14. The decipherment of Maya writing is detailed in Michael Coe, *Breaking the Maya Code* (London: Thames & Hudson, 2013).

15. Mayan numerals are vigesimally based, but some calendrical numerals use dots in the third position to represent 360 instead of 400, that is, they are a combination of base-20 and base-18 patterns. This so-called long-count system facilitated the specification of dates with respect to the creation of the universe in Mayan mythology.

16. This discussion of numerals only touches on a few of the ways in which numeral systems vary, ways that are particularly relevant for this book. For the most comprehensive and detailed look at the way numerals vary, see Stephen Chrisomalis, *Numerical Notation: A Comparative History* (New York: Cambridge University Press, 2010). Chrisomalis's work exhaustively categorizes numeral types according to a variety of functional parameters.

17. The single knot at the bottom of the cords, in the 'ones' position, represented different numbers in accordance with how many loops were needed to make it. In this way, it was clear that this position represented the "end" of the numeral. The remaining knots were simpler and occurred in clusters in the positions associated with particular exponents. The account I present here admittedly glosses over some of the complexity of this semiotic system, focusing on its decimal nature. For more on Incan numerals, see, for example, Gary Urton, "From Middle Horizon Cord-Keeping to the Rise of Inka Khipus in the Central Andes," *Antiquity* 88 (2014): 205–221.

18. Flegg, *Numbers through the Ages*.

3. A Numerical Journey around the World Today

1. The claim that Jarawara was anumeric was made in R. M. W. Dixon, *The Jarawara Language of Southern Amazonia* (Oxford: Oxford University Press, 2004), 559. I describe Jarawara numbers in Caleb Everett, "A Closer Look at a Supposedly Anumeric Language," *International Journal of American Linguistics* 78 (2012): 575–590, 583.

2. Cardinal number words like 'one,' 'two,' and 'three' describe sets of quantities, in contrast to ordinal words like 'first,' 'second,' and 'third.'

3. For more formal definitions of bases, see, for example, Bernard Comrie, "The Search for the Perfect Numeral System, with Particular Reference to Southeast Asia," *Linguistik Indonesia* 22 (2004): 137–145, or Harald Hammarström, "Rarities in Numeral Systems," in *Rethinking Universals: How Rarities Affect Linguistic Theory*, ed. Jan Wohlgemuth and Michael Cysouw (Berlin: De Gruyter Mouton, 2010), 11–59, 15, or Frans Plank, "Senary Summary So Far," *Linguistic Typology* 3 (2009): 337–345. Such formal definitions are avoided here as they differ from one another in minor ways that are not central to our story.

4. The frequency-based reduction of words is discussed, for instance, in Joan Bybee, *The Phonology of Language Use* (Cambridge: Cambridge University Press, 2001).

5. The finger basis of many spoken numbers is outlined in multiple works, including Alfred Majewicz, "Le Rôle du Doigt et de la Main et Leurs Désignations dans la Formation des Systèmes Particuliers de Numération et de Noms de Nombres dans Certaines Langues," in *La Main et les Doigts*, ed. F. de Sivers (Leuven, Belgium: Peeters, 1981), 193–212.

6. The numbers of languages in particular families are taken from M. Paul Lewis, Gary Simons, and Charles Fennig, eds., *Ethnologue: Languages of the World*, nineteenth edition (Dallas, TX: SIL International, 2016).

7. The word list and discussion of Indo-European forms is based on Robert Beekes, *Comparative Indo-European Linguistics: An Introduction* (Amsterdam: John Benjamins, 1995).

8. Andrea Bender and Sieghard Beller, " 'Fanciful' or Genuine? Bases and High Numerals in Polynesian Number Systems," *Journal of the Polynesian Society* 115 (2006): 7–46. See as well the discussion of Austronesian bases in Paul Sidwell, *The Austronesian Languages*, revised Edition (Canberra: Australian National University, 2013).

9. This insightful point was made by an anonymous reviewer.

10. Bernard Comrie, "Numeral Bases," in *The World Atlas of Language Structures Online*, ed. Matthew Dryer and Martin Haspelmath (Leipzig: Max Planck Institute for Evolutionary Anthropology, 2013), http://wals.info /chapter/131. For the most comprehensive survey of the world's verbal number systems, see the massive online database maintained by linguist Eugene Chan: https://mpi-lingweb.shh.mpg.de/numeral/.

11. This point is made in David Stampe, "Cardinal Number Systems," in *Papers from the Twelfth Regional Meeting, Chicago Linguistic Society* (Chicago: Chicago Linguistic Society, 1976), 594–609, 596.

12. Bernd Heine, *The Cognitive Foundations of Grammar* (Oxford: Oxford University Press 1997), 21.

13. For more details on the mechanics of number creation, see James Hurford, *Language and Number: Emergence of a Cognitive System* (Oxford: Blackwell, 1987).

14. The "basic numbers" referred to here are, defined pithily, cardinal terms used to describe the quantities of sets of items.

15. I am not the first to suggest that numbers serve as cognitive tools. This point has been advanced in several works, perhaps most clearly in Heike Wiese, "The Co-Evolution of Number Concepts and Counting Words,"

Lingua 117 (2007): 758–772, and Heike Wiese, *Numbers, Language, and the Human Mind* (Cambridge: Cambridge University Press, 2003).

16. The Indian merchant counting strategy is discussed in Georges Ifrah, *The Universal History of Numbers: From Prehistory to the Invention of the Computer* (London: Harville Press, 1998). It has also been suggested that base-60 strategies are due to a combination of decimal and base-6 systems, in which case they would still be partially based on human digits.

17. For an analysis of Oksapmin counting, see Geoffrey Saxe, "Developing Forms of Arithmetical Thought among the Oksapmin of Papua New Guinea," *Developmental Psychology* 18 (1982): 583–594. Counting among the Yupno is described in Jurg Wassman and Pierre Dasen, "Yupno Number System and Counting," *Journal of Cross-Cultural Psychology* 25 (1994): 78–94.

18. An overview of base-6 systems is given in Plank, "Senary Summary So Far." See also Mark Donohue, "Complexities with Restricted Numeral Systems," *Linguistic Typology* 12 (2008): 423–429, as well as Nicholas Evans, "Two *pus* One Makes Thirteen: Senary Numerals in the Morehead-Maro Region," *Linguistic Typology* 13 (2009): 321–335.

19. See Patience Epps, "Growing a Numeral System: The Historical Development of Numerals in an Amazonian Language Family," *Diachronica* 23 (2006): 259–288, 268.

20. These points are based in part on Hammarström, "Rarities in Numeral Systems," which surveys rare number bases in the world's languages.

21. Claims of the limits of numbers in Australian languages are made in Kenneth Hale, "Gaps in Grammar and Culture," in *Linguistics and Anthropology: In Honor of C. F. Voegelin*, ed. M. Dale Kinkade, Kenneth Hale, and Oswald Werner (Lisse: Peter de Ridder Press, 1975), 295–315, and R. M. W. Dixon, *The Languages of Australia* (Cambridge: Cambridge University Press, 1980). The detailed survey of Australian numbers discussed here is in Claire Bowern and Jason Zentz, "Diversity in the Numeral Systems of Australian Languages," *Anthropological Linguistics* 54 (2012): 133–160. Despite the relatively restricted number inventories of Australian languages, the majority of them also have grammatical means of expressing concepts like plural, singular, and even dual, meaning that their speakers frequently refer to discrete differences between smaller quantities though they have limited means of conveying minor discrepancies between larger quantities. Given that some Amazonian languages lack the latter sorts of grammatical means of encoding basic numerical concepts, and given that the most restricted number sys-

tems are found in Amazonian languages, it is fair to say that the most linguistically anumeric groups reside in Amazonia.

22. See Nicholas Evans and Stephen Levinson, "The Myth of Language Universals: Language Diversity and Its Importance for Cognitive Science," *Behavioral and Brain Sciences* 32 (2009): 429–448.

23. In this chapter we have discussed global patterns in cardinal numbers, words that describe the quantities of sets of items. The focus has been on the representation of words for positive integers, since other numbers (like fractions and negative numbers) are less common in the world's cultures and are also comparatively recent innovations. It is worth mentioning, though, that many generalizations we have highlighted also apply to fractions, given that these are based on integers in any given language. In English, for instance, fractions such as one tenth, one fifth, and so on, are inverted units taken from the basic decimal scale. This is not surprising, since it would be symbolically cumbersome to switch to, say, a senary base from a decimal one when speaking about fractions.

4. Beyond Number Words

1. See Matthew Dryer, "Coding of Nominal Plurality," in *The World Atlas of Language Structures Online,* ed. Matthew Dryer and Martin Haspelmath (Leipzig: Max Planck Institute for Evolutionary Anthropology, 2013), http://wals.info/chapter/33.

2. Stanislas Dehaene, *The Number Sense: How the Mind Creates Mathematics* (New York: Oxford University Press, 2011), 80.

3. Robert Dixon, *The Dyirbal Language of North Queensland* (New York: Cambridge University Press, 1972), 51.

4. Some morphological particulars in Kayardild are glossed over here. For more on the dual in this language, consult the following comprehensive grammatical description: Nicholas Evans, *A Grammar of Kayardild* (Berlin: Mouton de Gruyter, 1995), 184.

5. Greville Corbett, *Number* (Cambridge: Cambridge University Press, 2000), 20.

6. Wyn Laidig and Carol Laidig, "Larike Pronouns: Duals and Trials in a Central Moluccan Language," *Oceanic Linguistics* 29 (1990): 87–109, 92.

7. As an anonymous reviewer points out, some controversial claims of quadral markers, used in restricted contexts, have been made for the

Austronesian languages Tangga, Marshallese, and Sursurunga. See the discussion of these forms in Corbett, *Number,* 26–29. As Corbett notes in his comprehensive survey, the forms are probably best considered paucal markers. In fact, his impressive survey did not uncover any cases of quadral marking in the world's languages.

8. Boumaa Fijian grammatical number is discussed in R. M. W. Dixon, *A Grammar of Boumaa Fijian* (Chicago: University of Chicago Press, 1988).

9. Thomas Payne, *Describing Morphosyntax* (Cambridge: Cambridge University Press, 1997), 109.

10. Payne, *Describing Morphosyntax,* 98.

11. Payne, *Describing Morphosyntax.*

12. For a book-length discussion of grammatical number, see Corbett, *Number.*

13. Jon Ortiz de Urbina, *Parameters in the Grammar of Basque* (Providence, RI: Foris, 1989). Technically the verb agrees in number with the 'absolutive' noun, not the object, but this distinction is not important to our discussion.

14. Payne, *Describing Morphosyntax,* 108.

15. John Lucy, *Grammatical Categories and Cognition: A Case Study of the Linguistic Relativity Hypothesis* (Cambridge: Cambridge University Press, 1992), 54.

16. Caleb Everett, "Language Mediated Thought in 'Plural' Action Perception," in *Meaning, Form, and Body,* ed. Fey Parrill, Vera Tobin, and Mark Turner (Stanford, CA: CSLI 2010), 21–40. Note that the pattern described here is not the same as a verb agreeing with nominal number. The pattern in question is more similar to the *stampede* vs. *run* example, in which a verb has inherent plural connotations.

17. Dehaene, *The Number Sense.*

18. For evidence of the commonality of 1–3, see Frank Benford, "The Law of Anomalous Numbers," *Proceedings of the American Philosophical Society* 78 (1938): 551–572. For a discussion of the commonality of smaller quantities and of multiples of 10, see Dehaene, *The Number Sense,* 99–101.

19. This example of Roman numerals has been noted elsewhere, for instance, in Dehaene, *The Number Sense.*

20. The range of sounds in languages is taken from Peter Ladefoged and Ian Maddieson, *The Sounds of the World's Languages* (Hoboken, NJ: Wiley-

Blackwell, 1996). For one study on the potential environmental adaptations of languages, see Caleb Everett, Damián Blasi, and Seán Roberts, "Climate, Vocal Cords, and Tonal Languages: Connecting the Physiological and Geographic Dots," *Proceedings of the National Academy of Sciences USA* 112 (2015): 1322–1327.

5. Anumeric People Today

1. The Pirahã have been discussed extensively elsewhere, most notably in my father's book: Daniel Everett, *Don't Sleep, There Are Snakes: Life and Language in the Amazonian Jungle* (New York: Random House, 2008).

2. John Hemming, *Tree of Rivers: The Story of the Amazon* (London: Thames and Hudson, 2008), 181.

3. In fact, he became a very well-known scholar after encountering the Pirahã and has published numerous works on their language as well as other topics. These works have led to extensive discussion in academic circles, and in the media, on the nature of language. Most famously, perhaps, his research on the language suggests that the Pirahã language lacks recursion, a syntactic feature assumed by some linguists to occur in all languages.

4. These results on the imprecision of number-like words in the language are presented in Michael Frank, Daniel Everett, Evelina Fedorenko, and Edward Gibson, "Number as a Cognitive Technology: Evidence from Pirahã Language and Cognition," *Cognition* 108 (2008): 819–824. My discussion combines the results of the "increasing quantity elicitation" and "decreasing quantity elicitation" tasks in that study. The observation that all number-like words in the language are imprecise was offered earlier, in Daniel Everett, "Cultural Constraints on Grammar and Cognition in Pirahã: Another Look at the Design Features of Human Language," *Current Anthropology* 46 (2005): 621–646.

5. Pierre Pica, Cathy Lemer, Veronique Izard, and Stanislas Dehaene, "Exact and Approximate Arithmetic in an Amazonian Indigene Group," *Science* 306 (2004): 499–503.

6. Peter Gordon, "Numerical Cognition without Words: Evidence from Amazonia," *Science* 36 (2004): 496–499.

7. In other words, the correlation had what psychologists call a standard *coefficient of variation*. The coefficient of variation refers to the ratio one

arrives at by taking the standard deviation of responses and dividing it by the correct responses, for each target quantity. Gordon found that the coefficient of variation hovered around 0.15 for all quantities greater than three. We observed the same pattern in follow-up work among the Pirahã.

8. See Caleb Everett and Keren Madora, "Quantity Recognition among Speakers of an Anumeric Language," *Cognitive Science* 36 (2012): 130–141.

9. The results obtained at Xaagiopai do suggest that, when the Pirahã have had some practice with number words in their own language, they also begin to show signs of recognizing larger quantities more precisely. After all, their performance on the basic line matching task did seem to improve in that village after some number-word familiarization.

10. Interestingly, some languages in South Australia have "birth-order names," which indicate someone's relative age when contrasted to their siblings. As an anonymous reviewer points out, this is true in the Kaurna language, for example.

11. These Munduruku findings are presented in Pica et al., "Exact and Approximate Arithmetic in an Amazonian Indigene Group."

12. Pica et al., "Exact and Approximate Arithmetic in an Amazonian Indigene Group," 502.

13. Franc Marušič, Rok Žaucer, Vesna Plesničar, Tina Razboršek, Jessica Sullivan, and David Barner, "Does Grammatical Structure Speed Number Word Learning? Evidence from Learners of Dual and Non-Dual Dialects of Slovenian," *PLoS ONE* 11 (2016): e0159208. doi:10.1371/journal.pone.0159208.

14. Stanislas Dehaene, *The Number Sense: How the Mind Creates Mathematics* (New York: Oxford University Press, 2011), 264.

15. Koleen McCrink, Elizabeth Spelke, Stanislas Dehaene, and Pierre Pica, "Non-Developmental Halving in an Amazonian Indigene Group," *Developmental Science* 16 (2012): 451–462.

16. Maria de Hevia and Elizabeth Spelke, "Number-Space Mapping in Human Infants," *Psychological Science* 21 (2010): 653–660.

17. The study of the mental number line evident among the Munduruku is Stanislas Dehaene, Veronique Izard, Elizabeth Spelke, and Pierre Pica, "Log or Linear? Distinct Intuitions of the Number Scale in Western and Amazonian Indigene Cultures," *Science* 320 (2008): 1217–1220.

18. Rafael Núñez, Kensy Cooperrider, and Jurg Wassman, "Number Concepts without Number Lines in an Indigenous Group of Papua New Guinea," *PLoS ONE* 7 (2012): 1–8.

19. Elizabet Spaepen, Marie Coppola, Elizabeth Spelke, Susan Carey, and Susan Goldin-Meadow, "Number without a Language Model," *Proceedings of the National Academy of Sciences USA* 108 (2011): 3163–3168, 3167.

20. Only now are there signs that pressures from the outside will eventually yield the systematic adoption of numbers into these cultures. For instance, many governmental resources have recently been dedicated to familiarizing the Pirahã at Xaagiopai with Portuguese, including Portuguese number words.

6. Quantities in the Minds of Young Children

1. We do not know when exactly these number senses become accessible to us, though as we shall see, the approximate number sense is accessible at birth. My reference to number 'senses' owes itself to Stanislas Dehaene's fantastic book, *The Number Sense: How the Mind Creates Mathematics* (New York: Oxford University Press, 2011). As first noted in Chapter 4, the exact number sense is actually enabled by a more general capacity for tracking discrete objects. The quantitative function of this capacity is epiphenomenal. For mnemonic ease I refer to this quantitative function as the exact number sense, as it is what enables the relatively precise differentiation of smaller sets of items. For more on the general object-tracking or "parallel individuation" capacity that enables the discrimination of small quantities, see, for example, Elizabeth Brannon and Joonkoo Park, "Phylogeny and Ontogeny of Mathematical and Numerical Understanding," in *The Oxford Handbook of Numerical Cognition*, ed. Roy Cohen Kadosh and Ann Dowker (Oxford: Oxford University Press, 2015), 203–213.

2. One case for an innate language capacity is elegantly presented in Steven Pinker, *The Language Instinct: The New Science of Language and Mind* (London: Penguin Books, 1994). For more recent alternative perspectives, the reader may wish to consult accessible texts such as Vyv Evans, *The Language Myth: Why Language Is Not an Instinct* (Cambridge: Cambridge University Press, 2014) or Daniel Everett, *Language: The Cultural Tool* (New York: Random House, 2012).

3. Karen Wynn, "Addition and Subtraction by Human Infants," *Nature* 358 (1992): 749–750.

4. Furthermore, the study addressed some of the criticisms leveled at Wynn, "Addition and Subtraction by Human Infants," as well as other studies

that did not control for non-numerical confounds like amount, shape, and configuration of stimuli. See Fei Xu and Elizabeth Spelke, "Large Number Discrimination in 6-Month-Old Infants," *Cognition* 74 (2000): B1–B11.

5. I say "most infants" here, because for four of the sixteen infants who participated in the study, no staring differences were observed when they encountered novel amounts of dots.

6. Xu and Spelke, "Large Number Discrimination in 6-Month-Old Infants," B10.

7. This is an understandable issue with psychological research more generally, which is typically focused on peoples in Western, educated, and industrialized societies, since such peoples are easily accessible to most psychologists. See the discussion in Joseph Henrich, Steven Heine, and Ara Norenzayan, "The Weirdest People in the World?" *Behavioral and Brain Sciences* 33 (2010): 61–83.

8. The study described here is Veronique Izard, Coralie Sann, Elizabeth Spelke, and Arlette Streri, "Newborn Infants Perceive Abstract Numbers," *Proceedings of the National Academy of Sciences USA* 106 (2009): 10382–10385.

9. Such evidence does not suggest, however, that the human brain is *uniquely* hardwired for mathematical thought. As we will see in Chapter 7, other species also have an abstract number sense for differentiating quantities when the ratio between them is sufficiently large.

10. Jacques Mehler and Thomas Bever, "Cognitive Capacity of Very Young Children," *Science* 3797 (1967): 141–142. See also the enlightening discussion on this topic in Dehaene, *The Number Sense: How the Mind Creates Mathematics*, particularly as it relates to the work of Piaget. I should mention, however, that an insightful reviewer notes that there have been issues replicating the results of Mehler and Bever with very young children.

11. Kirsten Condry and Elizabeth Spelke, "The Development of Language and Abstract Concepts: The Case of Natural Number," *Journal of Experimental Psychology: General* 137 (2008): 22–38.

12. For a different perspective, see Rochel Gelman and C. Randy Gallistel, *Young Children's Understanding of Numbers* (Cambridge, MA: Harvard University Press, 1978), or Rochel Gelman and Brian Butterworth, "Number and Language: How Are They Related?" *Trends in Cognitive Sciences* 9 (2005): 6–10. Note that these works predate some of the research discussed here.

13. A more detailed discussion of the successor principle is presented in, for example, Barbara Sarnecka and Susan Carey, "How Counting Represents Number: What Children Must Learn and When They Learn It," *Cognition* 108 (2008): 662–674.

14. For more on the acquisition of these concepts by children in numerate cultures, I refer the reader to Susan Carey, *The Origin of Concepts* (Oxford: Oxford University Press, 2009), and Susan Carey, "Where Our Number Concepts Come From," *Journal of Philosophy* 106 (2009): 220–254.

15. See Elizabeth Gunderson, Elizabet Spaepen, Dominic Gibson, Susan Goldin-Meadow, and Susan Levine, "Gesture as a Window onto Children's Number Knowledge," *Cognition* 144 (2015): 14–28, 22.

16. See Barbara Sarnecka, Megan Goldman, and Emily Slusser, "How Counting Leads to Children's First Representations of Exact, Large Numbers," in *The Oxford Handbook of Numerical Cognition,* ed. Roy Cohen Kadosh and Ann Dowker (Oxford: Oxford University Press, 2015), 291–309. For more on the acquisition of one-to-one correspondence, see also Barbara Sarnecka and Charles Wright, "The Idea of an Exact Number: Children's Understanding of Cardinality and Equinumerosity," *Cognitive Science* 37 (2013): 1493–1506.

17. See Carey, *The Origin of Concepts.* Carey's account suggests that the innate exact differentiation of small quantities is the chief facilitator of the acquisition of other numerical concepts. In other words, the approximate number sense plays a less substantive role in the initial structuring of numbers, when contrasted to some other accounts. Some empirical support for her account is offered, for instance, in Mathiew Le Corre and Susan Carey, "One, Two, Three, Four, Nothing More: An Investigation of the Conceptual Sources of the Verbal Counting Principles," *Cognition* 105 (2007): 395–438. Debate remains among specialists as to how our innate number senses are fused. But it is generally agreed that both contribute to the eventual acquisition of numerical and arithmetical concepts.

18. The phrase "concepting labels" is taken from Nick Enfield, "Linguistic Categories and Their Utilities: The Case of Lao Landscape Terms," *Language Sciences* 30 (2008): 227–255, 253. For more on the way that number words serve as placeholders for concepts in the minds of kids, see Sarnecka, Goldman, and Slusser, "How Counting Leads to Children's First Representations of Exact, Large Numbers."

19. While truly representative cross-cultural studies on the development of numerical thought are largely missing in the literature, some recent work with a farming-foraging culture in the Bolivian rainforest, the Tsimane', explores these issues. The Tsimane' take about two to three times as long to learn to count, when contrasted with children in industrialized societies. See Steve Piantadosi, Julian Jara-Ettinger, and Edward Gibson, "Children's Learning of Number Words in an Indigenous Farming-Foraging Group," *Developmental Science* 17 (2014): 553–563. A very recent study of this group has found that their understanding of exact quantity correspondence correlates with knowledge of numbers and counting, as predicted by the account presented here. Interestingly, however, that same study suggests that there is at least one Tsimane' child "who cannot count but nevertheless understands the logic of exact equality." This is unexpected but not startling either. After all, we know that some humans (like number inventors) come to recognize exact equality without first counting. Of course, these Tsimane' kids still have exposure to counting and numerical semiotic practices, as they are embedded in a numerate culture. It is clear from all the relevant work, including that among the Tsimane', that learning to count greatly facilitates the subsequent recognition of precise quantities. See Julian Jara-Ettinger, Steve Piantadosi, Elizabeth S. Spelke, Roger Levy, and Edward Gibson, "Mastery of the Logic of Natural Numbers is not the Result of Mastery of Counting: Evidence form Late Counters," *Developmental Science* 19 (2016): 1–11. doi:10.1111/desc12459, 8.

7. Quantities in the Minds of Animals

1. For more on this experiment, of which I have provided only a basic summary, see Daniel Hanus, Natacha Mendes, Claudio Tennie, and Josep Call, "Comparing the Performances of Apes *(Gorilla gorilla, Pan troglodytes, Pongo pygmaeus)* and Human Children *(Homo sapiens)* in the Floating Peanut Task," *PLoS ONE* 6 (2011): e19555.

2. For evidence on the extent to which the collaboration between animals and humans impacted our species, see Pat Shipman, "The Animal Connection and Human Evolution," *Current Anthropology* 54 (2010): 519–538.

3. For more on Clever Hans, see Oscar Pfungst, *Clever Hans: (The Horse of Mr. von Osten) A Contribution to Animal and Human Psychology* (New York: Holt and Company, 1911).

4. See Charles Krebs, Rudy Boonstra, Stan Boutin, and A. R. E. Sinclair, "What Drives the 10-Year Cycle of Snowshoe Hares?" *Bioscience* 51 (2001): 25–35.

5. The emergence of prime numbers in such cycles is described in Paulo Campos, Viviane de Oliveira, Ronaldo Giro, and Douglas Galvão, "Emergence of Prime Numbers as the Result of Evolutionary Strategy," *Physical Review Letters* 93 (2004): 098107.

6. Nevertheless, it must be acknowledged that some invertebrate species exhibit behaviors consistent with rudimentary quantity approximation. See the survey in Christian Agrillo, "Numerical and Arithmetic Abilities in Non-Primate Species," in *Oxford Handbook of Numerical Cognition,* ed. Ann Dowker (Oxford: Oxford University Press, 2015), 214–236.

7. The numerical cognition of salamanders is described in Claudia Uller, Robert Jaeger, Gena Guidry, and Carolyn Martin, "Salamanders *(Plethodon cinereus)* Go for More: Rudiments of Number in an Amphibian," *Animal Cognition* 6 (2003): 105–112, and also in Paul Krusche, Claudia Uller, and Ursula Dicke, "Quantity Discrimination in Salamanders," *Journal of Experimental Biology* 213 (2010): 1822–1828. Results obtained with fish are described in Christian Agrillo, Laura Piffer, Angelo Bisazza, and Brian Butterworth, "Evidence for Two Numerical Systems That Are Similar in Humans and Guppies," *PLoS ONE* 7 (2012): e31923.

8. The seminal study of rats is that of John Platt and David Johnson, "Localization of Position within a Homogeneous Behavior Chain: Effects of Error Contingencies," *Learning and Motivation* 2 (1971): 386–414.

9. Regarding lionesses, see Karen McComb, Craig Packer, and Anne Pusey, "Roaring and Numerical Assessment in the Contests between Groups of Female Lions, *Panther leo,*" *Animal Behaviour* 47 (1994): 379–387. For findings on pigeons, see Jacky Emmerton, "Birds' Judgments of Number and Quantity," in *Avian Visual Cognition,* ed. Robert Cook (Boston: Comparative Cognition Press, 2001).

10. Agrillo, "Numerical and Arithmetic Abilities in Non-Primate Species," 217.

11. Results vis-à-vis dogs are offered in Rebecca West and Robert Young, "Do Domestic Dogs Show Any Evidence of Being Able to Count?" *Animal Cognition* 5 (2002): 183–186. For findings with robins, see Simon Hunt, Jason Low, and K. C. Burns, "Adaptive Numerical Competency in a Food-Hoarding

Songbird," *Proceedings of the Royal Society of London: Biological Sciences* 267 (2008): 2373–2379.

12. Agrillo et al., "Evidence for Two Numerical Systems That Are Similar in Humans and Guppies."

13. The similarity of the human and chimp genomes is described by The Chimpanzee Sequencing and Analysis Consortium, "Initial Sequence of the Chimpanzee Genome and Comparison with the Human Genome," *Nature* 437 (2005): 69–87. The value of genomic correspondence varies depending on the methods used, but is generally found to be greater than 95 percent. See also Roy Britten, "Divergence between Samples of Chimpanzee and Human DNA Sequences is 5% Counting Indels," *Proceedings of the National Academy of Sciences USA* 99 (2002): 13633–13635. For an exploration of the human genetic similarity to other species, visit http://ngm.nationalgeographic.com/2013/07/125-explore/shared-genes.

14. Mihaela Pertea and Steven Salzberg, "Between a Chicken and a Grape: Estimating the Number of Human Genes," *Genome Biology* 11 (2010): 206.

15. See Marc Hauser, Susan Carey, and Lilan Hauser, "Spontaneous Number Representation in Semi-Free Ranging Rhesus Monkeys," *Proceedings of the Royal Society of London: Biological Science* 267 (2000): 829–833. Some of Hauser's work has been called into question due to an inquiry conducted at Harvard, which found evidence that some of his results had been tampered with. The results in this particular study are not involved in that inquiry.

16. The results on this ascending task are described in Elizabeth Brannon and Herbert Terrace, "Ordering of the Numerosities 1–9 by Monkeys," *Science* 282 (1998): 746–749.

17. The chocolate experiment is described in Duane Rumbaugh, Sue Savage-Rumbaugh, and Mark Hegel, "Summation in the Chimpanzee *(Pan troglodytes),*" *Journal of Experimental Psychology: Animal Behaviors Processes* 13 (1987): 107–115.

18. Support for these claims is presented in Brannon and Terrace, "Ordering of the Numerosities 1–9 by Monkeys." With respect to baboons and squirrel monkeys, see Brian Smith, Alexander Piel, and Douglas Candland, "Numerity of a Socially Housed Hamadryas Baboon *(Papio hamadryas)* and a Socially Housed Squirrel Monkey *(Saimiri sciureus),*" *Journal of Comparative Psychology* 117 (2003): 217–225. For more on squirrel monkeys, see Anneke Olthof, Caron Iden, and William Roberts, "Judgements of Ordi-

nality and Summation of Number Symbols by Squirrel Monkeys *(Saimiri sciureus),"* Journal of Experimental Psychology: Animal Behaviors Processes 23 (1997): 325–339. Monkeys are capable of selecting the larger quantity of food items via approximation or via more exact methods that depend on training with numbers. Yet their quantity-discrimination skills are not restricted to the realm of consumables. Studies have also shown that rhesus monkeys can accurately choose the larger of two digital arrays of items presented via computer screen, even after non-numeric properties, such as surface area of the presented stimuli, are controlled. See Michael Beran, Bonnie Perdue, and Theodore Evans, "Monkey Mathematical Abilities," in *Oxford Handbook of Numerical Cognition,* ed. Ann Dowker (Oxford: Oxford University Press, 2015), 237–259.

19. The cross-species evidence for an exact number sense, enabled by what is often referred to as the parallel individuation system, is weaker and, to some researchers, marginal at best. See discussion in Beran, Perdue, and Evans, "Monkey Mathematical Abilities." Researchers have not fully fleshed out the range of similarity between our innate number senses and those evident in other species, such as our primate relatives.

20. Elizabeth Brannon and Joonkoo Park, "Phylogeny and Ontogeny of Mathematical and Numerical Understanding," in *Oxford Handbook of Numerical Cognition,* ed. Ann Dowker (Oxford: Oxford University Press, 2015), 209.

21. Irene Pepperberg, "Further Evidence for Addition and Numerical Competence by a Grey Parrot *(Psittacus erithacus)," Animal Cognition* 15 (2012): 711–717. For results with Sheba, see Sarah Boysen and Gary Berntson, "Numerical Competence in a Chimpanzee *(Pan troglodytes)," Journal of Comparative Psychology* 103 (1989): 23–31.

22. Pepperberg, "Further Evidence for Addition and Numerical Competence by a Grey Parrot *(Psittacus erithacus),"* 711.

8. Inventing Numbers and Arithmetic

1. To read more about how patterns in language impact thought, see Caleb Everett, *Linguistic Relativity: Evidence across Languages and Cognitive Domains* (Berlin: De Gruyter Mouton, 2013).

2. James Hurford, *Language and Number: Emergence of a Cognitive System* (Oxford: Blackwell, 1987), 13. The perspective I present here is influenced by the

more recent work of Heike Wiese, "The Co-Evolution of Number Concepts and Counting Words," *Lingua* 117 (2007): 758–772. She observes on page 762 that "the dual status of counting words crucially means that they are numbers (as well as words), rather than number names, that is, they do not refer to extra-linguistic 'numbers', but instead are used as numbers right away." Wiese also notes that the traditional "numbers-as-names" approach overlooks ordinal ('first,' 'second,' etc.) and nominal (e.g., "the #9 bus") number words.

3. Karenleigh Overmann, "Numerosity Structures the Expression of Quantity in Lexical Numbers and Grammatical Number," *Current Anthropology* 56 (2015): 638–653, 639. For a reply to this article, see Caleb Everett, "Lexical and Grammatical Number Are Cognitive and Historically Dissociable," *Current Anthropology* 57 (2016): 351.

4. Stanislas Dehaene, *The Number Sense: How the Mind Creates Mathematics* (New York: Oxford University Press, 2011), 80.

5. See Kevin Zhou and Claire Bowern, "Quantifying Uncertainty in the Phylogenetics of Australian Number Systems," *Proceedings of the Royal Society B: Biological Sciences* 282 (2015): 2015–1278. These findings are consistent with the related discussion of Australian numbers in Chapter 3, which was based on a separate study—one also co-authored by Bowern.

6. The physical bases of number words has been observed in many sources, for instance, in Bernd Heine, *Cognitive Foundations of Grammar* (Oxford: Oxford University Press, 1997).

7. Apart from any particular contestable details of this account, little doubt remains that number words are verbal tools, not merely labels for concepts that all people are innately predisposed to recognize. See also Wiese, "The Co-Evolution of Number Concepts and Counting Words," 769, where she notes, for example, that "counting words are verbal instances of numerical tools, that is, verbal tools we use in number assignments."

8. There are many works on embodied cognition. For one extensive survey of this topic, consult Lawrence Shapiro (ed.), *The Routledge Handbook of Embodied Cognition* (New York: Routledge, 2014). In contrast to the account presented here, some archaeologists have focused on how body-external features have impacted the innovation of numbers. See, for example, Karenleigh Overmann, "Material Scaffolds in Numbers and Time," *Cambridge Archaeological Journal* 23 (2013): 19–39. They suggest an alternate account, according to which materials like beads, tokens, and tally marks served as material placeholders for concepts that were then instantiated linguistically. No doubt such

artifacts, like other material factors, placed additional pressures on humans to invent and refine numbers. (See Chapter 10.) But the perspective espoused here is that the anatomical pathways to numbers are more basic ontogenetically and historically when contrasted to any other (no doubt extant) external numeric placeholders. Fingers are, after all, more experientially primal than such body-external material stimuli. In addition, there is a clear tie between numeric language and the body (see Chapter 3), which suggests the primacy of the body in inventing numbers, not just labeling them after material placeholders for numbers are invented. The claim here is not, however, that material technologies and symbols do not also play a role in fostering numerical thought, and the research of such archaeologists is crucial to elucidating the extent of that role. As humans engaged with numbers materially, we no doubt faced greater pressures to extend our number systems in new ways. But, even considering such pressures, our fingers are what enabled the very invention of numbers, at least in most cases.

9. Rafael Núñez and Tyler Marghetis, "Cognitive Linguistics and the Concept(s) of Number," in *The Oxford Handbook of Numerical Cognition,* ed. Roy Cohen Kadosh and Ann Dowker (Oxford: Oxford University Press, 2015), 377–401, 377.

10. For a detailed consideration of the role of metaphors in the creation of math, see George Lakoff and Rafael Núñez, *Where Mathematics Comes From: How the Embodied Mind Brings Mathematics into Being* (New York: Basic Books, 2001). For a more recent consideration, see Núñez and Marghetis, "Cognitive Linguistics and the Concept(s) of Number."

11. Núñez and Marghetis, "Cognitive Linguistics and the Concept(s) of Number," 402.

12. Núñez and Marghetis, "Cognitive Linguistics and the Concept(s) of Number," 402.

13. Of course, kids are frequently counting actual objects when they learn and use math. Yet the larger point is that in all contexts, including abstract ones, we use a physical grounding to talk about how we mentally manipulate the quantities represented through numbers. Such metaphorical bases of numerical language are common throughout the world. In Chapter 5 it was noted, though, that number lines are not used in all cultures to make sense of quantities.

14. The value of gestures in exploring human cognition is evident, for example, in Susan Goldin-Meadow, *The Resilience of Language: What Gesture*

Creation in Deaf Children Can Tell Us about How All Children Learn Language (New York: Psychology Press, 2003). The findings on mathematical gestures discussed here are also taken from Núñez and Marghetis, "Cognitive Linguistics and the Concept(s) of Number."

15. These points on brain imaging are adapted from Stanislas Dehaene, Elizabeth Spelke, Ritta Stanescu, Philippe Pinel, and Susanna Tsivkin, "Sources of Mathematical Thinking: Behavioral and Brain-Imaging Evidence," *Science* 284 (1999): 970–974. The spatial interference example is adapted from Dehaene, *The Number Sense: How the Mind Creates Mathematics,* 243.

16. This SNARC effect was first described in Stanislas Dehaene, Serge Bossini, and Pascal Giraux, "The Mental Representation of Parity and Number Magnitude," *Journal of Experimental Psychology: General* 122 (1993): 371–396.

17. See Heike Wiese, *Numbers, Language, and the Human Mind* (Cambridge: Cambridge University Press, 2003), and Wiese, "The Co-Evolution of Number Concepts and Counting Words," for a detailed account of how syntax may impact numerical thought. According to Wiese, this sort of linguistically based thinking enables us to use not just cardinal numbers, which refer to the values of particular sets of items, but also ordinal and nominal numbers. (See note 2.) Such valuable insights should not be overextended either. The range of diversity in the world's languages should give us pause before concluding that syntactic influences play a major role in the expansion of numerical thought in all cultures. Considering the extent to which some languages allow so-called free word order and do not have rigid syntactic constraints like English, such caution is prudent. These include many languages with rich case systems that convey who the subject and object are irrespective of their position in a clause (Latin, for instance). The speakers of some languages with freer syntax still acquire numbers. This does not imply that syntax does not play a role in facilitating our own acquisition of such concepts. However, any influence of grammar on the way we learn numbers likely varies substantially across cultures.

18. For more on brain-to-body size ratios, see Lori Marino, "A Comparison of Encephalization between Ondontocete Cetaceans and Anthropoid Primates," *Brain, Behavior and Evolution* 51 (1998) 230–238. For further details of the human cortex, see Suzana Herculano-Houzel, "The Human Brain in Numbers: A Linearly Scaled-Up Primate Brain," *Frontiers in Human Neu-*

roscience 3 (2009): doi:10.3389/neuro.09.031.2009. The neuron count used here is taken from Dorte Pelvig, Henning Pakkenberg, Anette Stark, and Bente Pakkenberg, "Neocortical Glial Cell Numbers in Human Brains," *Neurobiology of Aging* 29 (2008): 1754–1762.

19. IPS activation in monkeys is described in Andreas Nieder and Earl Miller, "A Parieto-Frontal Network for Visual Numerical Information in the Monkey," *Proceedings of the National Academy of Sciences USA* 19 (2004): 7457–7462. The interaction of cortical regions and particular quantities has been discussed in various works, including Dehaene, *The Number Sense: How the Mind Creates Mathematics*, 248–251.

20. Relevant locations in the IPS are presented in Stanislas Dehaene, Manuela Piazza, Philippe Pinel, and Laurent Cohen, "Three Parietal Circuits for Number Processing," *Cognitive Neuropsychology* 20 (2003): 487–506. Degree of activation is discussed in Philippe Pinel, Stanislas Dehaene, D. Rivière, and Denis LeBihan, "Modulation of Parietal Activation by Semantic Distance in a Number Comparison Task," *Neuroimage* 14 (2001): 1013–1026.

21. See Dehaene, *The Number Sense: How the Mind Creates Mathematics*, 241, for imaging evidence of the verbal expansion of quantitative reasoning. Given that the hIPS is clearly associated with numerical cognition, some researchers have posited a brain "module" dedicated to numerical thought. See Brian Butterworth, *The Mathematical Brain* (London: Macmillan, 1999). It is important to recall that the cortex is highly plastic and that, although certain parts of the brain may be associated with certain functions, these regions may vary across individuals.

9. Numbers and Culture

1. Khufu was about 8 meters taller before its outer shell eroded. Using the original height ($139 + 8$), we have $147 \times 2 \times \pi = 924$, while the perimeter is $230 \times 4 = 920$.

2. The most widely cited survey of color terms is Brent Berlin and Paul Kay, *Basic Color Terms: Their Universality and Evolution* (Berkeley: University of California Press, 1969). Fascinating data on the cross-cultural variability of olfactory categorizations are presented in Asifa Majid and Niclas Burenhult, "Odors are Expressable in Language, as Long as You Speak the Right Language," *Cognition* 130 (2014): 266–270.

3. The correlation between numbers and subsistence strategy is presented in the global survey in Patience Epps, Claire Bowern, Cynthia Hansen, Jane Hill, and Jason Zentz, "On Numeral Complexity in Hunter-Gatherer Languages," *Linguistic Typology* 16 (2012): 41–109. The findings on Bardi are taken from the same work, p. 50.

4. As we saw in Chapter 8, however, some Australian languages do have a number word for 5, which leads to the relatively rapid innovation of larger numbers.

5. For more on the isolation of some Amazonian groups, see Dylan Kesler and Robert Walker, "Geographic Distribution of Isolated Indigenous Societies in Amazonia and the Efficacy of Indigenous Territories," *PLoS ONE* 10 (2015): e0125113.

6. Although we should not denigrate particular linguistic and cultural traditions, we can avoid such prejudices while simultaneously acknowledging that numerical technologies enable certain types of reasoning that, in turn, yield new kinds of innovations. These innovations, it should be admitted, ultimately include such benefits as medicinal technologies that yield longer life spans. So even though numbers may not lead to impartially considered "better" or "more advanced" lives, they were indubitably crucial to the transition to longer life spans. Of course numbers were also crucial to less pleasant developments, such as mechanized warfare.

7. See, for instance, Andrea Bender and Sieghard Beller, "Mangarevan Invention of Binary Steps for Easier Calculation," *Proceedings of the National Academy of Sciences USA* 111 (2014): 1322–1327, as well as Andrea Bender and Sieghard Beller, "Numeral Classifiers and Counting Systems in Polynesian and Micronesian Languages: Common Roots and Cultural Adaptations," *Oceanic Linguistics* 25 (2006): 380–403. See also Sieghard Beller and Andrea Bender, "The Limits of Counting: Numerical Cognition between Evolution and Culture," *Science* 319 (2008): 213–215.

8. For birth-order names in South Australian languages, see Rob Amery, Vincent Buckskin, and Vincent "Jack" Kanya, "A Comparison of Traditional Kaurna Kinship Patterns with Those Used in Contemporary Nunga English," *Australian Aboriginal Studies* 1 (2012): 49–62.

9. Bender and Beller, "Mangarevan Invention of Binary Steps for Easier Calculation," 1324.

10. For more on the potential advantages of such technologies, consult, for example, Michael Frank, "Cross-Cultural Differences in Representations

and Routines for Exact Number," *Language Documentation and Conservation* 5 (2012): 219–238. See also the survey of technologies like abaci in Karl Menninger, *Number Words and Number Symbols* (Cambridge, MA: MIT Press, 1969).

11. The recent rediscovery of the eastern hemisphere's oldest zero, in Cambodia, is described in Amir Aczel, *Finding Zero: A Mathematician's Odyssey to Uncover the Origins of Numbers* (New York: Palgrave Macmillan, 2015). Given the heavy influence of Indian culture on the Khmer, it is assumed that zero was transferred from India to Cambodia. Still, the oldest definitive instance of zero in the Old World is that found near Angkor, first discovered in the 1930s and rediscovered in 2015 by Aczel—who scoured many stone stelae to find it.

12. For rich surveys of the world's written numeral systems, see Stephen Chrisomalis, *Numerical Notation: A Comparative History* (New York: Cambridge University Press, 2010), as well as Stephen Chrisomalis, "A Cognitive Typology for Numerical Notation," *Cambridge Archaeological Journal* 14 (2004): 37–52.

13. There is some argument as to whether Egyptian hieroglyphs were innovated independently of an awareness of writing in Sumeria. They appear on the scene not long after the development of Mesopotamian writing, by most accounts. Given that Sumeria and Egypt are relatively proximate geographically, it is likely that Egyptians developed hieroglyphs only after they became knowledgeable of the existence of writing.

14. For a look at early cuneiform, see Eleanor Robson, *Mathematics in Ancient Iraq: A Social History* (Princeton, NJ: Princeton University Press, 2008). For a discussion of numbers in early written forms, see Stephen Chrisomalis, "The Origins and Co-Evolution of Literacy and Numeracy," in *The Cambridge Handbook of Literacy*, ed. David Olson and Nancy Torrance (New York: Cambridge University Press, 2009), 59–74. Chrisomalis describes the copresence of numerals and ancient writing systems, though he notes that this copresence may be coincidental.

15. However, I should be clear that tally systems do not necessarily develop into writing systems or written numerals. The Jarawara tally system, pictured in Figure 2.2, did not eventually yield a native Jarawara system of writing. The same could be said of some tally systems that have existed in Africa and elsewhere for thousands of years. But even though the existence of a tally system may not be a sufficient condition for the invention of

writing, it may increase the likelihood of a writing system being innovated.

10. Transformative Tools

1. The effects of climatic shifts on human speciation are discussed in Susanne Shulz and Mark Maslin, "Early Human Speciation, Brain Expansion and Dispersal Influenced by African Climate Pulses," *PLoS ONE* 8 (2013): e76750. On the potential influence of Toba, see Michael Petraglia, "The Toba Volcanic Super-Eruption of 74,000 Years Ago: Climate Change, Environments, and Evolving Humans," *Quaternary International* 258 (2012): 1–4. On the advantages of coastal southern Africa during this time frame, see Curtis Marean, Miryam Bar-Matthews, Jocelyn Bernatchez, Erich Fisher, Paul Goldberg, Andy Herries, Zenobia Jacobs, Antonieta Jerardino, Panagiotis Karkanas, Tom Minichillo, Peter Nilssen, Erin Thompson, Ian Watts, and Hope Williams, "Early Human Use of Marine Resources and Pigment in South Africa during the Middle Pleistocene," *Nature* 449 (2007): 905–908.

2. The tempered stone tools in question present advantages when contrasted to the Oldowan and Acheulean stone tools that persevered in the human lineage for about 2.5 million years, beginning about 2.6 million years ago. See, for instance, Nicholas Toth and Kathy Schick, "The Oldowan: The Tool Making of Early Hominins and Chimpanzees Compared," *Annual Review of Anthropology* 38 (2009): 289–305.

3. For more on the Blombos Cave finds see, for example, Christopher Henshilwood, Francesco d'Errico, Karen van Niekerk, Yvan Coquinot, Zenobia Jacobs, Stein-Erik Lauritzen, Michel Menu, and Renata Garcia-Moreno, "A 100,000-Year-Old Ochre Processing Workshop at Blombos Cave, South Africa," *Science* 334 (2011): 219–222.

4. Francesco d'Errico, Christopher Henshilwood, Marian Vanhaeren, and Karen van Niekerk, "*Nassarius krausianus* Shell Beads from Blombos Cave: Evidence for Symbolic Behaviour in the Middle Stone Age," *Journal of Human Evolution* 48 (2005): 3–24, 10.

5. See Susan Carey, "Précis of the Origin of Concepts," *Behavioral and Brain Sciences,* 34 (2011): 113–167, 159. Carey's point is offered in response to Karenleigh Overmann, Thomas Wynn, and Frederick Coolidge, "The Prehistory of Number Concepts," *Behavioral and Brain Sciences* 34 (2011):

142–144. The authors of that piece suggest that the beads at Blombos may have served as actual material numbers since "a string of beads possesses inherent characteristics that are also components of natural number" (p. 143). In other words they suggest the beads *were* the first numbers, and that numbers were first material and became linguistic after people labeled the material numbers. It seems more plausible that such valuable homogeneous items created *pressures* for the innovation of linguistic numbers, a creation only made possible because of human anatomical characteristics. For instance, Overmann, Wynn, and Coolidge note that "a true numeral list emerges when people attach labels to the various placeholder beads" (p. 144). Such an account glosses over the less speculative psycholinguistic evidence (see Chapter 5) demonstrating that human adults cannot consistently discriminate quantities of things like beads without first using numbers. I believe the account also underappreciates the linguistic data demonstrating that people name numbers after hands or fingers, not after things like beads. In short, our hands serve as the true gateway to numbers, even if body-external items like beads create pressures for their creation.

6. The survey demonstrating a correlation between population size and religion is presented in Frans Roes and Michel Raymond, "Belief in Moralizing Gods," *Evolution and Human Behavior* 24 (2003): 126–135. My comments here are based partially on Ara Norenzayan and Azim Shariff, "The Origin and Evolution of Religious Prosociality," *Science* 322 (2008): 58–62. The advantages of within-group cooperation for cultural adaptive fitness, enhanced by religion, are discussed in Scott Atran and Joseph Henrich, "The Evolution of Religion: How Cognitive By-Products, Adaptive Learning Heuristics, Ritual Displays, and Group Competition Generate Deep Commitments to Prosocial Religions," *Biological Theory* 5 (2010): 18–130.

7. Greek, Hebrew, Arabic, and other languages associated with the major religions in question have decimal-based number systems. Therefore, the pattern being highlighted here is likely a by-product of linguistic decimal systems. Regardless, the pattern is also fundamentally due to the structure of the human hands. This point merits attention, I think, since the profundity ascribed to some religious numbers is not commonly recognized to be influenced in any manner by human anatomy.

8. Which is not to suggest that all spiritually significant numbers are neatly divisible by ten. In fact, some smaller ones are prime numbers: there is the three of the holy trinity or the seven deadly sins or the seven virtues

of the holy spirit or the seven days of creation. Note that all these numbers are less than ten. Even exceptions greater than ten are not always as exceptional as they may seem. Consider the importance of twelve to Islam, Judaism, and Christianity: the twelve Imams, the twelve tribes of Israel, and the twelve apostles. As noted in Chapter 3, duodecimal bases also have potential manual origins as well.

9. A critical look at P values and their history is presented in Regina Nuzzo, "Scientific Method: Statistical Errors," *Nature* 506 (2014): 150–152.

ACKNOWLEDGMENTS

This book was made possible in part by a generous award from the Carnegie Corporation of New York. The statements made and views expressed are solely my own, of course.

I am very thankful for the invaluable guidance of my editor at Harvard University Press, Jeff Dean, and also appreciative of Michael Fisher, who first saw promise in the work. The book benefited tremendously from the insightful comments offered by four knowledgeable reviewers. I am grateful to each of them for taking the time to read the initial manuscript. Their comments and criticisms simply made the book better. There are too many brilliant scholars directly or indirectly associated with the research described in this book to offer adequate thanks. If you are one of those people doing the fascinating work on which this book is based, thanks for doing what you do.

Some of this book was written aboard the *MV Explorer*, on Semester at Sea, during long ocean passages made more enjoyable by the people I met on that ship. Some of it was also written at Lagoa da Conceição, ensconced in a paradisiacal hillside. Most of it was written at the University of Miami, a wonderful place to write and do research. I only work there because, some years ago, faculty in the Department of Anthropology took a chance on a young scholar they first interviewed during a layover at Miami International Airport. I remain grateful to them for that. Thanks are due as well to my other colleagues at UM, who have made my experience here so pleasant. I have also been privileged to have many great students at UM, with whom I have discussed some of the ideas in this book.

Both of my parents influenced this work, directly and indirectly, in ways that are hopefully evident in its pages. I thank them for that, and for everything else they have given me. Much of which, I realize, I don't even remember. I will also always be grateful for my wonderful sisters and their amazing families, and for the Scottis as well. Finally, this book would not have been possible without my wife Jamie and our son Jude.

INDEX